SCIENCE AND TECHNOLOGY EDUCATION
AND FUTURE HUMAN NEEDS

Volume 7

Energy Resources
in Sci

Science and Technology Education and Future Human Needs

General Editor: JOHN LEWIS
Malvern College, United Kingdom

Vol. 1 LEWIS & KELLY
Science and Technology Education and Future Human Needs

Vol. 2 FRAZER & KORNHAUSER
Ethics and Social Responsibility in Science Education

Vol. 3 WADDINGTON
Education, Industry and Technology

Vol. 4 GRAVES
Land, Water and Mineral Resources in Science Education

Vol. 5 KELLY & LEWIS
Education and Health

Vol. 6 RAO
Food, Agriculture and Education

Vol. 7 KIRWAN
Energy Resources in Science Education

Vol. 8 BAEZ
The Environment and Science and Technology Education

Vol. 9 TAYLOR
Science Education and Information Transfer

Energy Resources in Science Education

Edited by

D. F. KIRWAN
University of Rhode Island, U.S.A.

Published for the

ICSU PRESS

by

PERGAMON PRESS
OXFORD · NEW YORK · BEIJING · FRANKFURT
SÃO PAULO · SYDNEY · TOKYO · TORONTO

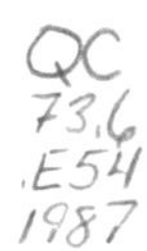

U.K.	Pergamon Press, Headington Hill Hall, Oxford OX3 0BW, England
U.S.A.	Pergamon Press, Maxwell House, Fairview Park, Elmsford, New York 10523, U.S.A.
PEOPLE'S REPUBLIC OF CHINA	Pergamon Press, Room 4037, Qianmen Hotel, Beijing, People's Republic of China
FEDERAL REPUBLIC OF GERMANY	Pergamon Press, Hammerweg 6, D-6242 Kronberg, Federal Republic of Germany
BRAZIL	Pergamon Editora, Rua Eça de Queiros, 346, CEP 04011, Paraiso, São Paulo, Brazil
AUSTRALIA	Pergamon Press Australia, P.O. Box 544, Potts Point, N.S.W. 2011, Australia
JAPAN	Pergamon Press, 8th Floor, Matsuoka Central Building, 1-7-1 Nishishinjuku, Shinjuku-ku, Tokyo 160, Japan
CANADA	Pergamon Press Canada, Suite No. 271, 253 College Street, Toronto, Ontario, Canada M5T 1R5

First edition 1987

Library of Congress Cataloging in Publication Data

Energy resources in science education.
(Science and technology education and future human needs; vol. 7)
Based on the papers presented at the International Conference on Science and Technology Education and Future Human Needs, held in Bangalore, India, Aug. 6–14, 1985.
1. Force and energy—Study and teaching—Congresses.
2. Power resources—Study and teaching—Congresses.
I. Kirwan, D. F. (Donald F.) II. Bangalore Conference on Science and Technology Education and Future Human Needs (1985) III. Series.
QC73.6.E54 1987 333.79'07'1 87–2394

British Library Cataloguing in Publication Data

Energy resources in science education.
(Science and technology education and future human needs; v. 7).
1. Power resources—Study and teaching
I. Kirwan, D.F. II. Series
333.79'07 TJ163.2

ISBN 0–08–033950–6 (Hardcover)
ISBN 0–08–033951–4 (Flexicover)

Printed in Great Britain by A. Wheaton & Co. Ltd., Exeter

Foreword

The Bangalore Conference on "Science and Technology Education and Future Human Needs" was the result of extensive work over several years by the Committee on the Teaching of Science of the International Council of Scientific Unions. The Committee received considerable support from Unesco and the United Nations University, as well as a number of generous funding agencies.

Educational conferences have often concentrated on particular disciplines. The starting point at this Conference was those topics already identified as the most significant for development, namely Health; Food and Agriculture; Energy; Land, Water and Mineral Resources; Industry and Technology; the Environment; Information Transfer. Teams worked on each of these, examining the implications for education at all levels (primary, secondary, tertiary, adult and community education). The emphasis was on identifying techniques and resource material to give practical help to teachers in all countries in order to raise standards of education in those topics essential for development. As well as the topics listed above, there is also one concerned with the educational aspects of Ethics and Social Responsibility. The outcome of the Conference is this series of books, which can be used for follow-up meetings in each of the regions of the world and which can provide the basis for further development.

JOHN L. LEWIS
Secretary, ICSU-CTS

Editor's Preface

The material contained in this volume is the result of the combined efforts of the planners, hosts and participants at the 1985 International Conference on Science and Technology Education and Future Human Needs which was held in Bangalore, India on 7–14 August 1985. The planning group spent innumerable hours outlining the activity schedule, topical areas of concern, and other items relating to the formal structure and operation of the Conference. The hosts ensured that the work of the Conference proceeded as smoothly as possible, attending to the special needs of the logistical, organisational, comfort and leisure needs of the organisers and participants alike. The participants of the Energy Group worked long hard hours preparing materials for this book. The team facilitators: Ian Winter, Tim Hickson, John Fowler and Vivian Talisayon deserve special recognition due to their additional efforts to organise and edit the incoming written (often by hand) materials. Final copy for Section C of this book underwent final rewrite and edit by Tim Hickson. I am indeed grateful for his help, assistance and willingness to undertake that task. Editorial assistance was performed by Patricia Kirwan, Cynthia Yemma and Gayle Ater. The whole Bangalore Conference could not have been a success without the efforts of Peter Kelly, Chairman of ICSU-CTS and in particular of John Lewis, Secretary of ICSU-CTS who handled each and every small detail of the Conference in addition to being the group leaders' shepherd.

Acknowledgments

The energy group at Bangalore benefited from the materials which were brought by the participants for sharing with their colleagues. Several individuals, associations and agencies also sent sample energy educational materials developed in their country which ranged from complete energy activity packages to sample lesson plans. It is this latter group that the participants at Bangalore wish to give special formal recognition by an acknowledgment of their contribution to the success of the Conference. Materials were received from:

Dr. Hann, Ann Chin
Korean Elementary Science Education Association
Department of Science Education
Inchon Teachers College
Inchon, Korea

Poon Poh Kong
Director
SEAMEO Regional Centre for Education in
Science & Mathematics
Glugor, Penang
Malaysia

Dr. M. Labor
Department of Physics
Fourah Bay College
University of Sierra Leone
Freetown, Sierra Leone

Professor P. E. Spargo
Director, Science Education Unit
University of Cape Town
Rondebosch 7700
South Africa

Winston K. King
Lecturer in Science Education
University of the West Indies
School of Education
Cave Hill, Barbados

Robert C. P. Westbury
Executive Director
S.E.E.D.E. Foundation
Edmonton, Alberta
Canada

Dr. Reinders Duit
Institute for Science Education (IPN)
University of Kiel
Oshausentrabe 40
D 23 Kiel 1
Federal Republic of Germany

Kenneth Young
Department of Physics
The Chinese University of Hong Kong

Koichi Saito
Komaba Toho High School Teacher
Komaba Toho School
4–5–1 Ikejiri Setagaya-ky
Tokyo 154, Japan

Thomas Varghese
Secretary General
ZASE
Zambia Association for Science Education
P.O. Box RW–335
Ridgeway
Lusaka, Zambia

Andrew Clegg
Head, Dept. of Math. & Science Education
University of Botswana

Prof. Khadim Ali Hashmi
Science Education Section
Scientific Society of Pakistan
Curriculum Research & Development
Wahdat Colony, Lahore 16, Pakistan

John Hartman
Atomic Energy of Canada Limited
Whiteshell Nuclear Research Establishment
Pinawa, Manitoba R0E 1 L0

Contents

List of Contributors

D. R. Baluragi	Jawaharnagar High School, Raichur, India
J. Dunin-Borkowski	al Niepodliglosci, Warsaw, Poland
A. P. French	Massachusetts Institute of Technology, Cambridge, Massachusetts, USA
J. M. Fowler	NSTA, Washington DC, USA
Hanna Goldring	Weizmann Institute of Science, Rehovet, Israel
M. P. Govindarajan	Chalavara High School, Kerala, India
H. Harnaes	University of Oslo, Oslo, Norway
K. Heiskanen	Karjalankata 12 Heinola, Finland
T. D. R. Hickson	The King's School, Worcester, UK
Changfeng Jiang	No. 4 High School, Beijing, China
Y. B. Kamble	Community Science Centre, Ahmedabad, India
D. F. Kirwan	University of Rhode Island, USA
M. Korek	Rowdoh High School, Beirut, Lebanon
Joanna Kumi	Ghana Academy of Arts and Science, Accra, Ghana
B. G. Kusuma	Acharya Pathasala Girls High School, Bangalore, India
J. L. Lewis	Malvern College, Malvern, UK
E. Lisk	Sierra Leone Grammar School, Freetown, Sierra Leone
Rose Malone	Trinity College, Dublin 2, Ireland
C. W. S. Murthy	J. M. Institute of Technology, Chitradurga, India
M. H. Padakannaya	Gambia High School, Banjul, The Gambia
Antonella Prat	IRRSAE, Pieminte, Italy
M. V. Rajasekharan	Asian Institute for Rural Development, Bangalore, India
Jose A. Rodriguez	CENAMEC, Caracas, Venezuela
Chris Shea	Ipswich State High School, Queensland, Australia
T. P. Sukumaran	High School, Kerala, India

T. Subahan	University Kebangsaan, Bangi, Selangor, Malaysia
R. Supornpaibul	Chulalongkorn University Demonstration School, Bangkok, Thailand
V. M. Talisayon	Institute for Science and Mathematics, Quezon City, Philippines
I. Winter	Claremont High School, Tasmania, Australia

Introduction

At no other time in the history of the world has there been a greater need for a scientifically literate citizenry. This is the age of uncertain world energy supplies, social disruptions, economic problems, overpopulation, and outright starvation of a significant portion of the population. This is also the age of an unprecedented rate of scientific and technological advancement. The energy factor plays an important causal role in the continuation of many of these problems and achievements.

With an adequate scientific knowledge base and an awareness of the varied and complex linkages between energy and environmental, political, economic, health and social issues, tomorrow's adults will be better equipped to participate in helping solve their nation's energy problems. This may be through the participation in decision-making processes or through some energy conversion technological breakthrough.

Teachers will play a significant role in ultimately solving energy problems. Attitudes along with technology are important in working towards solutions. Studies have shown that attitudes and ideas instilled in childhood can easily become habits. It is likely that the students entering our primary schools will be the generation to first feel the full impact of the approaching era of energy shortages. It is important that they, unlike a large segment of today's adult population, respond in a rational way based upon a realistic appreciation of the many and often related factors which are now beginning to govern national energy policies and the search for and development of alternate energy sources.

The teachers' responsibility is a critical one in teaching about the sources and uses of energy, the multifaceted energy problem and the energy conservation and environmental preservation ethic. It is vital that teachers who will be involved in teaching energy-related materials do so based upon a good scientific knowledge base and an awareness of what constitutes good teaching practice.

The purpose of this book is to disseminate some ideas about energy and teaching practices that are believed to have contributed to successful teaching of energy in the schools around the world.

Most of the "how to teach" suggestions included herein are widely practised. However, the innovative development of and/or successful implementation of many of these practices should be of interest to the teachers who will read this volume. The differences in most of these techniques and strategies of teaching are primarily associated with differences in national or local settings and flavours and in national and local priorities. For example, in many parts of the world, the only way to teach energy topics is to use them to enhance and illustrate the science content of existing academic courses within a curriculum.

Various parts of this book were written by the participants in the Conference on Science and Technology Education and Future Human Needs which was held in Bangalore, India, from 6 to 14 August 1985. The Energy Group had as its prime consideration, the development of materials which could serve as examples of effective ways energy could be taught in the schools. Consequently, this book is organised in such a way that it may be of most use to practising teachers and/or school administrators who may be interested in incorporating energy topics into their school's courses/curriculum.

SECTION A

Energy and Education

Introduction

The beginning of this volume related to energy education concerns itself with material related to the general ideas about what energy is, the understanding of energy conversion, and the energy perspective in education. In the past few years there has been a dramatic shift in thought relative to the conceptualisation of energy. Professor Dunin-Borkowski presents a rather disturbing (for those who feel comfortable with the status quo) and enlightening (for those who realise that energy has evolved into an interdisciplinary concept) discourse on the concept of energy based on its universality. Professor French's article, which helps clarify many of the physics concepts related to energy conversion, together with several articles on the teaching of energy (ranging from energy program development to energy teaching methodology), complete the general energy education material developed for the Bangalore Conference.

1

The Concept of Energy, Its Structure and Teaching Strategy

J. DUNIN-BORKOWSKI

University of Warsaw, Poland

Scientific concepts usually have their source in common life. One of the rare exceptions is the concept of energy. This concept emerged at the beginning of the eighteenth century (Bernoulli in 1717, or perhaps Kepler in 1609),[1]* and then was gradually developed during the next centuries.

Energy, of course, exists only as a theoretical concept; it does not exist physically like atoms or molecules do.[2] Nevertheless, the concept has undergone a curious materialisation as we talk about the "lack", "saving", "buying", "wasting", or "flowing" of energy. Should we then, as Schmidt does,[3] treat it as a sort of substance? Or, as Feynman introduces the concept of energy,[4] stress its abstractness, treating it as a constant number?

The practical importance of the concept of energy need not be argued. Newspapers and other media are full of headlines such as "Energy is a Quality of Life", "Energy Resources are Diminishing", "Save Energy". Even those who do not remember too much from school physics are aware that "energy crisis" means trouble with heating or fuel supply; "an energy price rise" means higher expenses. In these ways, a general perception of energy is formed. This vague, intuitive picture is seldom satisfactory.[5] A proper understanding of energy problems is needed to influence constructively the development of a modern society.

Energy has grown into an interdisciplinary concept and a tool for description and analysis of phenomena in every domain without exception. The universality of the energy concept is the source of its exceptional importance but, on the other hand, it creates difficulties by its formation.

* Superscript numbers refer to Notes at the end of the chapter.

Physical quantities can be grouped in several classes:

1. Quantities that are uniquely determined by some attribute of a certain material object or phenomenon, e.g. mass, velocity.
2. Quantities that depend on more than one attribute of a material object or phenomenon, e.g. momentum. A majestically stalking elephant and an artillery missile may have the same momentum although they seem to be quite different at face value. Such quantities are usually related to some conservation law.
3. Quantities that depend not only on the various attributes of an object or phenomenon, but also on the circumstances. Such a quantity is energy. Is this the only case?

When we start to explain what energy is, the explanation depends on the particular domain with which we are dealing. Up to now, physicists have been of the opinion that the only way of introducing the energy concept is to derive it from the definition of work. This is simply following the historical path. In 1891 Maxwell wrote: "energy is the capacity of doing work", although even at that time some people were aware that the energy concept is far broader. In 1848 Rankine described energy as the capacity of making change. But for the decades that followed, the work metaphor continued to function. In physics textbooks one could read about "capacity of doing work" or "stored work". At its birth the concept of energy emerged from mechanics and later was extended to heat phenomena. Today, energy plays a crucial role not only in physics but in many other places as well. The diversity of forms and circumstances in which energy now appears does not allow one to connect energy only with work.

During recent decades, as the whole complexity of the energy concept and the corresponding educational difficulties have been recognised, new projects have been designed in which work is no longer the starting point for energy, e.g. Nuffield O-level Physics. These projects, however, have placed emphasis on the difficult concept of energy transformation.

Shortly thereafter it was claimed that science teaching should aim not only at intellectual development but at the preparation for action and social life (Lewis, Science in Society Project).[6]

It is easy to see that there is discrepancy in the understanding of energy in disciplines other than physics and the sense it is given in traditional schooling. Its meaning is still different in common life. The energy transformations in chemical reactions or in living organisms and biological systems are seldom connected with performance of work. For a typical family of four, whose annual consumption of energy is about 300 GJ, at most 30 per cent is related with work. This share drops to smaller percentage in those countries which are less motorised. In the consciousness of a common user the linkage of energy with work is then very loose.

Energy is so complex that it is not possible to give a comprehensible definition and explain all of its features in a single step. The full characteristics of energy consist of a broad spectrum of interrelated properties. The manner and depth of interpretation both change and develop, and depend on the level and the needs of the user. In the hot polemical discussions currently flourishing in journals and conferences, different approaches to teaching energy are argued, and it is very difficult to choose between the opponents. As shown by Jon Ogborn, several different approaches are logically equivalent.[7] The adversaries are arbitrarily choosing different starting points.

It is obvious that the choice of a starting point and teaching strategy is essential for educational achievement. Different approaches may have different weaknesses. Moreover, since it is not possible to explore in general education the entire realm of energy, it follows that the emphasis should be given to those topics which are most important in today's society.

Let us see the structure of the energy concept. As was mentioned, it is a disjunctive concept like a multi-headed dragon.

Ten Truths About Energy

1. *Energy is a physical quantity.*
   It follows that energy comparisons, estimates, and measurements should be possible.

2. *Energy is a function of state. The absolute value of energy is unknown.*
   This explains the tendency to materialise the concept of energy.

3. *Energy appears in various forms which differ both formally and phenomenologically.*
   Contrary to other physical quantities, because energy appears in phenomena which differ by their nature and mechanism, the mathematical formulas describing different forms of energy are different.

4. *Energy can change from one form to another in a chain of transformations.*
   Physical phenomena are related to energy transformations.

5. *Energy transformation can occur by conversion of form or by the transfer of energy between the elements of the system.*

6. *Processes of energy transfer are:*
   (a) *Work*
   (b) *Heat*
   (c) *Electric work*
   (d) *Radiation*

In the first three of these processes, it is possible to measure the energy transfer.

7. *Transformations of energy are one way processes.*
All processes in nature are irreversible. Therefore the full description of events requires using not only the First but also the Second Law of Thermodynamics. In using the energy concept one must take into account the direction of the process. The direction of the process can be expressed descriptively by such terms as "energy degradation", "value of the energy", and "primary energy". Using fuel as a starting point for the development of the energy concept, as suggested by Rogers, is an example. In the papers of Lewis, Ogborn, Timbal, Duclaux and others the necessity of showing directionality has been stressed.[8] It should be noted that in considering the transformation of mechanical energy, e.g. in elastic deformation of a spring, only the change of Gibbs free energy is taken into account.[9]

8. *Transformation of energy depends on the converter used.*
Chain branching can occur in the converter.
The branching of the energy chain leads to the idea of converter efficiency.

9. *The total energy involved in the transformation chain is conserved.*
While the principle of conservation of energy is underlined emphatically in traditional teaching, the relationship between energy and the Second Law of Thermodynamics is commonly disregarded. This practice leads to a contradiction with common experience.

10. *New forms of energy can be defined:*

    (a) As a quantity which may be needed to satisfy the conservation law.
    (b) As a calculational methodology for the energy of microscopic interactions in a previously constructed model.

Assuming that these "truths" properly characterise the whole energy concept spectrum, they can serve as the basis for comparison and analysis of different teaching approaches. Such a study allows the development of teaching strategies designed to attain preset goals. Let us consider some examples:

Energy as Stored Work

In this case the change of energy occurs as a result of performed work. The system must be isolated so other forms of energy transfer can be neglected. This situation is very peculiar and seldom met in practice. The energy introduced in this way has distinct features of a substance; the generalisation of the energy concept is difficult.

Energy as a Capacity for Doing Work

In this case we have a specific situation: an adiabatic energy transformation. This time energy is not a substance but a condition of the system such that work can be performed. In this case generalisation is once again rather difficult.

Energy as Warming Capability

In this case there is a transfer of heat energy from a hot to a cold body. Energy again looks like a substance which can flow and cannot be destroyed. Again it is a special case but closer to reality. The direction and irreversibility are very distinct. Such approaches can be found in Hungarian and Danish projects (Kedves, Veje—GIREP, 1981).

Energy in the Transformation Chain

Instead of looking for a special case which is then gradually extended to the more general situation, the picture of energy flow through the chain of transformations is given at the beginning. Energy is not defined but is "something" which changes from one link to another following the chain of events. The French project is an example here.[10] The criticism of such an approach is that the concept of energy is vague and gives no indication of how to measure it.

Primary Energy

We avoid the artificial situation by building up the energy concept from the beginning: primary energy at the beginning of the chain. According to Rogers: fuel is a primary source of energy and the starting point for many different chains of transformations. In this way energy has a practical sense and the direction of changes is stressed.

Energy Converters

The use of an energy converter can determine the direction of the process. In this case the law of energy conservation can be applied and the change of energy allows one to estimate the value of energy transformed. In some cases where work is performed, work can be a measure of the energy transformed by the converter. It is attained not by choosing an artificial situation but a special device (Timbal-Duclaux—GIREP, 1981).

This article has attempted to show how energy evolves from something strictly connected with work to a more general concept, one closer to real life experience.

References

1. Brouzeng, P., Etude historique de la notion d'energie, *Bull. Un. des Phys.* **74,** 1135 (1980).
2. Warren, J. W., Energy and its carriers: a critical analysis, *Phys. Educ.* **18,** 213 (1983).
3. Schmid, G. B., Energy and its carriers, *Phys. Educ.* **17,** 212 (1982).
4. Feynman, *et al., The Feynman Lectures on Physics.*
5. Barbetta, M. G. *et al.,* An investigation of students' frameworks about motion and the concepts of force and energy, *Proc. GIREP Conf.,* 1984, p. 219, Utrecht (1985).
6. Lewis, J. L., Energy education and society, *Proc. GIREP Conf.,* 1981, p. 285, Budapest (1981).
7. Ogborn, J., Teaching about energy, *Proc. GIREP Conf.,* 1981, Budapest (1981).
8. GIREP Conf., 1981, Budapest (1981).
9. Barrat, J. P. and Guignier, G., L'energie potentielle: energie interne ou energie libre?, *Bull. Un. des Phys.* **75,** 115 (1980).
10. Agabra, J. *et al., Science Physiques, 3 Colleges,* Collection Libres Parcours, Clasique Hachette (1980).

2

Physics Concepts in Energy Conversion

A. P. FRENCH

Massachusetts Institute of Technology, USA

There is one single principle that dominates the entire subject of energy conversion in physics: *Energy is conserved*. Because of that fact, energy is one of the most important concepts and quantities in the physicists's description of Nature. We cannot be sure that conservation of energy is absolutely universal. However, it does apply to every situation or process known to us at the present time. On those occasions in the past when it seemed that energy conservation might have failed, further investigation has always led to the discovery of some other form of energy, or of some previously unknown agency by which energy could be carried away. Thus energy conversion and energy conservation are tightly connected to one another.

Energy Conversion and Conservation in Mechanics

Simple mechanics gives us the initial basis for conservation of energy, and also for relating energy to work:

1. Observation shows that, if an object acquires speed v in falling from rest through a vertical distance h near the Earth's surface, then v^2 is proportional to h. In modern terms we say that its initial potential energy (PE) mgh, has been converted into kinetic energy (KE) $\frac{1}{2}mv^2$. Why the factor of ½ in the expression for KE? Because we can then *construct* an energy conservation statement. If we just overcome gravity by an upward force equal to mg, this force does work mgh in raising an object through h. If we let gravity act alone, it does an equal amount of work *mgh on the object* in bringing it down again. But also, by Newton's Second Law, we know that the object accelerates in such a way that $gh = \frac{1}{2}v^2$. Thus, if we *define* kinetic energy as $\frac{1}{2}mv^2$, and *define* potential energy as mgh, we can write, for a freely falling object,

Initial KE = 0 Final KE = ½ $mv^2 = mgh$
Initial PE = mgh Final PE = 0

Hence,

$$\text{Initial KE} + \text{Initial PE} = \text{Final KE} + \text{Final PE}.$$

Moreover, potential energy is being converted to kinetic energy throughout the descent, and at any intermediate point we have

$$E = KE + PE = mgh = \text{constant}.$$

Thus we develop the notion of the *total* mechanical energy, E, of the object, and this is conserved in free fall (if air resistance is negligible).

2. The above example is a very special case. If an object is attached to a string, so that it is forced to move in a circular arc as a pendulum, we find that, if it descends along the arc through a vertical distance, h, starting from rest, we again have a gain of KE equal to mgh. The tension in the string, acting always at right angles to the displacement, does no work, but it does change the direction of the motion. The same holds for any effectively frictionless track along which the object descends. We can in every such case write:

$$KE + \text{Gravitational PE} = E_{total} = \text{constant}.$$

This embodies a fundamental feature: *Energy is a scalar quantity;* kinetic energy depends only on the size of the velocity, not on its direction.

3. Energy conservation is mechanics can take various forms, but it always applies to a complete system. In examples (1) and (2), above, the system is basically the earth and the object of mass m; the PE is the mutual gravitational potential energy of m and the Earth. Energy is converted from potential to kinetic, or conversely. The motion of a mass attached to a spring involves similar energy conversions, although in this case the PE is that associated with elastic deformation of the spring. But there may be systems in which, for practical purposes, the energy is all kinetic and is shared in different ways between the parts of a system, as in the elastic collision between two objects, for which we can write:

$$(KE_1 + KE_2)_{initial} = (KE_1 + KE_2)_{final}$$

During the collision, KE is briefly converted into elastic PE, but is then converted back again.

4. One very important form of mechanical energy is rotational KE. For a rigid rotating object, the rotational energy $\frac{1}{2} I\omega^2$ can represent a form of energy storage (without any linear motion) which can be converted into other forms. One example is the conversion of rotational energy to electrical energy in an electric generator.

Basic Units of Energy

From the basic definitions of PE and KE in mechanics, it follows that mechanical energy is equivalent to work and can be measured in terms of force × distance. Using the Système International (SI) we have as the unit of energy the joule (J):

$$1\ \mathrm{J} \equiv 1\ \mathrm{N.m} \equiv 1\ \mathrm{kg.m^2/s^2}.$$

The joule is then the fundamental SI unit for all forms of energy. However, as a matter of history and/or convenience other units are still widely used for forms of energy other than mechanical; we shall draw attention to these where appropriate.

For convenience in describing smaller or larger amounts of energy over a wide range, the usual prefixes micro (10^{-6}), milli (10^{-3}), kilo (10^{3}), mega (10^{6}), etc., are used to indicate submultiples or multiples of the joule (or of other units of energy). A table of conversions between the joule and other units of energy is given at the end of this paper (Table 1).

Heat and Thermal Energy

1. Ever since the development of the kinetic theory of matter, it has been recognised that heat and thermal energy are simply randomised forms of mechanical energy. However, the description of exchanges and conversions of thermal energy is dominated by the concept of temperature. It is not surprising that the degree Kelvin (K) is one of the seven SI base units. Transfers of heat are often most usefully described in terms of the temperature changes they produce, according to the basic equation of calorimetry:

$$H = m.C.\triangle T,$$

where H is the amount of heat defined in terms of the temperature change $\triangle T$ produced in a mass m of a substance of specific heat C. From this equation we have the definition of the practical unit of heat, the Calorie as the amount of thermal energy to raise 1 kg of water through 1 K. (This unit is often called the kilogram-calorie, to distinguish it from the original small calorie, defined as the amount of heat to raise 1 g of water through 1 K. Note that the large calorie is the one used in describing the calorific values of foods.)

It was the experiments of J. P. Joule that proved the existence of a precise equivalence in the conversion of mechanical work into heat. In 1948 the joule was officially designated to replace the Calorie as the unit of heat:

$$1\ \mathrm{Cal} = 4184\ \mathrm{J}$$
$$1\ \mathrm{J} = 2.39 \times 10^{-4}\ \mathrm{Cal}$$

2. One of the most important features of energy conversions involving heat is the existence of fundamental limitations on such conversions, as described by

the Second Law of Thermodynamics. The conversion of heat into work always involves the degradation of thermal energy, that is the taking of thermal energy (H_1) from a reservoir at one temperature (T_1) and the returning of part of that energy (H_2) to a second reservoir at a lower temperature (T_2). In this process, the amount of energy that can be extracted as useful work (W) is strictly limited according to the relationship

$$W/H_1 \leq (T_1 - T_2) / T_1$$

where T is the absolute temperature. This means that high efficiency in any power plant for converting heat into work depends on having the hot reservoir at as high a temperature as possible. The cold reservoir will often have approximately the temperature of the environment (or actually be the environment).

In the reverse process, of converting work into heat, the efficiency is greatest when the temperature difference is least:

$$H_2 / W \leq T_2 / (T_1 - T_2).$$

This maximum efficiency (or coefficient of performance) can be considerably greater than unity, a fact that is exploited in the use of heat pumps to heat houses in winter, using the house as the hot reservoir and the ground as the cold reservoir. Air conditioners and refrigerators also, of course, are forms of heat pumps.

Electric and Magnetic Energy

1. Because unlike charges attract one another, work must be done to produce a separation of positive from negative charge; this can result in a storage of energy. A typical practical example is a charged capacitor; the amount of energy stored is given by the familiar formulas:

$$W = \tfrac{1}{2} CV^2 = \tfrac{1}{2} QV = Q^2/2C$$

where C is the capacitance, V the potential difference between the plates, and Q the amount of charge on the capacitor plates. This energy can be considered as stored in the electric field between the plates. If C is in farads (F), V in volts (V), and Q in coulombs (C), then the energy W is in joules. Thus, for example, a capacitor of 2×10^{-6} F charged to a potential difference of 1000 V has 1 J of stored energy, the equivalent of raising 1 kg through about 10 cm at the Earth's surface, or of heating 1 g of water through about 0.25 K.

2. It is a consequence of Faraday's law of induction (coupled with Lenz's law) that work must be done to establish a current in a circuit containing inductance (i.e. in practice, any circuit). The work can be considered as being done to push the charge carriers in the current against the back EMF associated with the changing magnitude of the current. The result is an amount of stored energy given by:

$$W = \tfrac{1}{2} LI^2$$

where I is the current and L is the self-inductance of the circuit. This energy can be considered as stored in the magnetic field. If L is in henries (H) and I in amperes (A), then W is in joules. Thus a storage of 1 J would be obtained, for example, with a current of 10 A in an inductance of 20 mH.

3. In a circuit containing inductance and capacitance (i.e. any circuit) the total stored energy E is given by:

$$E = \tfrac{1}{2} LI^2 + Q^2/2C.$$

Since the current I is equal in magnitude to dQ/dt, this equation can be rewritten:

$$E = \tfrac{1}{2} L(dQ/dt)^2 + Q^2/2C.$$

This equation then has precisely the same form as that for mechanical energy conservation for a mass on a spring:

$$E = \tfrac{1}{2} m(dx/dt) + \tfrac{1}{2} kx^2$$

where k is the spring constant. (It is worth noting that the elastic stored energy in a spring is itself electrostatic energy, resulting from pushing or pulling the atoms in the spring away from their normal equilibrium positions. Electrostatic forces must be overcome to do this.) It follows that there can be a periodic interconversion between the electric and magnetic energies in the circuit, just as between the potential and kinetic energies of the mechanical system. The mechanical system oscillates with a period $2\pi\,(m/k)^{1/2}$; the electrical system with a period $2\pi\,(LC)^{1/2}$.

4. Since any electrical circuit also contains resistance, there is always a conversion (loss) of electromagnetic energy over to thermal energy dissipated in the resistor. Thus in a simple LCR circuit, if we ignore energy lost by electromagnetic radiation, the initial stored energy ($\tfrac{1}{2} LI^2 + \tfrac{1}{2} CV^2$) will all ultimately be deposited in the resistor, which of course becomes hot in consequence.

Radiant Energy

1. By this we shall mean *electromagnetic* radiation. Sound, of course, can also be regarded as radiation in a medium, but its source is a mechanical pressure or displacement of the same kind; energy *conversion* in a basic sense is not involved.

2. Electromagnetic radiation, classically considered, is the energy of associated electric and magnetic fields, traveling at a characteristic speed which in general depends on the frequency or wavelength but which in vacuum is the speed of light, c. The source of electromagnetic radiation is electric charges undergoing acceleration or deceleration; some of the kinetic energy of

the charges is converted into the energy of the electromagnetic field. There is no upper or lower limit, so far as we know, to the wavelength or frequency range of the electromagnetic spectrum; all forms of such radiation, from radio waves of many kilometers wavelength down to gamma rays of 10^{-14} m wavelength, fit into this same description.

3. From the standpoint of quantum physics, the emission or absorption of radiation is a matter of the creation or annihilation of individual photons of energy hf, where h is Planck's constant and f the classical frequency of the radiation. Energy conservation then requires that the energy needed or released is directly related to a corresponding change in the energy state of an atom, a free electron, a nucleus, or some other particle. The photon description makes appropriate the use of atomic or nuclear unit of energy, the electron-volts (eV) or multiples thereof:

$$1 \text{ eV} = 1.602 \times 10^{-19} \text{ J}$$
$$1 \text{ keV} = 1.602 \times 10^{-16} \text{ J}$$
$$1 \text{ MeV} = 1.602 \times 10^{-13} \text{ J}$$

The electron-volt can properly be considered a natural atomic unit of energy, since it represents the energy acquired by one elementary charge falling through 1 volt, a potential difference typical (in order of magnitude) of chemical cells whose EMF is the physical expression of individual processes of atomic ionisation within the cell.

4. Processes involving the absorption, rather than the emission, of radiation are of very great importance. For us earth-dwellers the supreme example is, of course, the arrival of radiant energy from the Sun, without which we could not exist. The ultimate conversion of this radiant energy to heat keeps the oceans from freezing, and the process of photosynthesis underlies most of our food supply. In terms of basic physics, the fundamental processes are those of photon absorption leading to a change of energy state for an electron, an atom, or a molecule, as exploited in photoelectric devices, solar cells, etc.

Nuclear Energy

1. The forms of energy we have described so far fall into three basic categories: (a) mechanical kinetic energy, (b) gravitational potential energy and (c) electromagnetic energy (which includes the energy associated with all forms of contact interaction between atoms). To these must be added nuclear energy, which is an expression of the forces of interaction within the nucleus, primarily between neutrons and protons. Nuclear energy can be released as a result of spontaneous nuclear decays or of artificially induced nuclear reactions. Some examples are:

(i) Alpha decay:

$$_{84}\text{Po}^{210} \rightarrow {}_{82}\text{Pb}^{206} + {}_{2}\text{He}^{4} + 5.30 \text{ MeV}$$

(ii) Beta decay:

$$ {}_0n^1 \rightarrow {}_1H^1 + e^- + \bar{\nu}_e + 0.78\ \text{MeV} $$

$$ {}_{15}P^{32} \rightarrow {}_{16}S^{32} + e^- + \bar{\nu}_e + 1.70\ \text{MeV} $$

(iii) Slow neutron radiative capture:

$$ {}_0n^1 + {}_{45}Rh^{103} \rightarrow {}_{45}Rh^{104} + \gamma + 6.8\ \text{MeV} $$

(iv) Charged-particle fusion reaction:

$$ {}_1H^2 + {}_1H^2 \rightarrow {}_1H^3 + {}_1H^1 + 4.0\ \text{MeV} $$

(v) Neutron-induced fission:

$$ {}_0n^1 + {}_{92}U^{235} \rightarrow {}_{56}Ba^{141} + {}_{36}Kr^{92} + 3\,{}_0n^1 + 175\ \text{MeV} $$

The details of the individual processes are not important for our present purposes; what *is* relevant is that each of them represents a process in which nuclear energy is converted into the kinetic energy of the emitted particle or particles. (In process (iii) the emitted particle is a photon.)

2. The beta-decay process (ii) is of particular interest, since historically it was one of those cases in which it appeared that the principle of conservation of energy might fail. The neutrino was simply *postulated* in 1930 to save the conservation principle, but because of its almost negligible interaction with matter was not discovered experimentally until 1956.

Mass-Energy

1. When Einstein in 1905 proposed his famous equation $E = mc^2$, he provided a unifying principle for all processes of energy transfer or conversion, regardless of the particular forces or interactions involved:

ENERGY IS MASS

$$ 1\ \text{J} = 1.113 \times 10^{-17}\ \text{kg} $$

$$ 1\ \text{kg} = 8.988 \times 10^{16}\ \text{J} $$

2. Direct calculational use of Einstein's mass-energy equivalence tends to be limited to nuclear processes, because only there are the fractional mass changes of the particles big enough (of the order of 0.1 per cent) to be directly measurable. This makes it possible to predict the energy release in a nuclear reaction from a prior knowledge of the isotopic masses involved, as obtained from mass spectrometry. It permits one to say that surplus energy could in principle be released through fusion reactions involving a total of less than 56 nucleons (neutrons and protons) or through fission processes resulting in two fragments, each with more than 56 nucleons. This is because the nucleus of Fe^{56} is more tightly bound, i.e. has less mass per nucleon, than any other nucleus.

3. However, it is important to realise that *any* change in the energy of a body implies a corresponding change of its mass. A ball in motion has more mass than the same ball at rest. The heated filament of a lamp has more mass than the same filament when cold. A charged capacitor has more mass than the same capacitor uncharged. The reactants in an exothermic chemical reaction in which heat and light escape from the system have a greater total mass than the products. In all such cases, it is erroneous to say that mass is *converted* into energy or *vice versa*; mass *is* energy, and in a closed system this mass-energy is constant. Einstein combined the separate laws of conservation of mass and conservation of energy into a single law that underlies all the energy conversion processes we have discussed.

Power

1. For practical purposes, the *rate* of doing work or of transferring or converting energy may be of great importance. Therefore, although it is not, strictly speaking, a basic physical concept, the measure of *power* plays a prominent role in the description of energy-conversion processes. The SI unit of power is the watt:

$$1 \text{ watt (W)} = 1 \text{ J/s}$$

2. Power measured in watts appears most frequently, of course, in connection with electrical systems. For the simplest case of a resistive circuit, with voltage and current in phase, we have the familiar results:

$$P = VI = I^2R = V^2/R$$

The conversion of electrical energy into heat or mechanical work is, of course, the basis of a large fraction of all industrial and domestic machines or appliances. The reverse conversion of mechanical to electrical energy is the task of electric generators.

The custom of using instantaneous power to characterise electrical energy transfers has led to the use of units of energy defined by power × time:

$$1 \text{ kilowatt-hour (kWh)} = 3.6 \times 10^6 \text{ J}$$

Other such units are sometimes useful in particular contexts. For example, the megawatt-day (8.6×10^{10} J) is useful as a unit for describing the energy output of a generating station or the energy consumption of a town.

The Varieties of Energy Conversion

The preceding sections have mentioned only a few specific examples of energy storage and conversion. In practice the number of possibilities is almost limitless. In Table 2 we show, with the help of a rectangular display, a wider selection of such possibilities. This certainly lays no claim to completeness, and the reader may well be able to find entries for some of the blank rectangles (or conversely, types of energy conversion not representable in this scheme) and will probably wish to supplement the existing entries in many places.

A particularly beautiful case in which many different kinds of energy conversion are involved is provided by the life history of a star. The process begins with gas and dust at very low density in space. Gravitational forces begin to draw the material together, and gravitational potential energy is converted to kinetic energy. When a certain degree of contraction has occurred, the density is big enough to lead to very frequent collisions; the result is to randomise the motions, and the kinetic energy is transferred into heat and a small fraction into radiation, by which the contracting cloud is made visible. In the process the originally neutral atoms become ionised, and the centre of the star becomes a fully ionised plasma. When the central temperature becomes high enough the processes of nuclear fusion can begin. The collisions between the nuclei become sufficiently energetic to overcome the Coulomb repulsion. Successive nuclear reactions have the effect of converting hydrogen to helium with the release of gamma radiation and a great flood of neutrinos. The neutrinos are able to escape almost unimpeded, but the gamma radiation interacts strongly with nuclei and electrons as it travels outward, and in the process the mean energy per photon is reduced from the order of MeV to a few eV, without, however, any net absorption of energy taking place. Thus the final result is that all the nuclear energy released within a given time at the centre of the star escapes into space from the star's surface, most of it (in the case of our sun, at any rate) as visible or near visible (UV or IR) electromagnetic radiation. The subsequent history of a star after it has "burnt" its hydrogen at the centre involves other fascinating energy conversions, but the details at this stage may depend markedly on particular features, notably the mass of the star and its angular momentum.

Some Comparative Magnitudes

It is useful, and very instructive, to have some sense of the relative amounts of energy involved in typical energy conversions or energy-storage mechanisms. Below are a few arbitrarily chosen examples. Each reader of this paper should construct such a list, based on experiences with which he or she is familiar. The list given here (rounded off to orders of magnitude in joules) has deliberately been kept within the range of energies on the human scale of activities:

Some Typical Energy Values

1 J — Work to pick up an apple (1 N through 1 m!).
Acoustic energy emitted in a 1-hour lecture.

10 J — Rotational KE of wheels of bicycle travelling at 20 mph.
Electrostatic energy stored in a 20μF capacitor charged to 1000 V (e.g. for high-intensity photo-flash unit).

10^2 J	KE of baseball traveling at 100 mph. Total KE of the molecules in 1 litre of air at room temperature.
10^3 J	1 Btu (energy to heat 1 lb of water through 1 °F). Gravitational PE gained by a proficient high-jumper. Electrical energy needed per second to drive a 1 hp motor.
10^4 J	Chemical energy stored in a flashlight dry cell. Electromagnetic radiant energy in 1 cubic mile of sunlight (at earth's distance from sun).
10^5 J	Energy (mgh) expended by a person ascending through 150 m. Energy used in one cycle of operation of an electric toaster.
10^6 J	KE of a car traveling at 100 km/hr. Nutritional energy in a full glass of milk. Chemical energy stored in a car battery.
10^7 J	Basic human energy requirement per day. Thermal energy needed for a hot bath.
10^8 J	Energy from combustion of 1 gallon of gasoline or kerosene. Solar energy received per m^2 per day at earth.
10^9 J	Heat of combustion of 1000 ft^3 of natural gas. KE of Boeing 727 (mass about 80 tons) flying at 600 km/hr.
10^{10} J	Energy from burning 1 ton of wood. Energy equivalent of 2 barrels of oil (bbl).
10^{11} J	1 Megawatt-day. Lifetime nutritional energy needed per human being. Energy from fission of 1 g of uranium atoms. Mass-energy equivalent of 1 mg of matter.

TABLE 1
Energy Units—a Conversion Table

		eV	MeV	amu*	erg	joule	ft.lb	Cal†	kWh	MW.day	g	kg
1 ev	=	1	10^{-6}	1.1×10^{-9}	1.6×10^{-12}	1.6×10^{-19}	1.2×10^{-19}	3.8×10^{-23}	4.5×10^{-26}	1.9×10^{-30}	1.8×10^{-33}	1.8×10^{-36}
1 MeV	=	10^{6}	1	1.1×10^{-3}	1.6×10^{-6}	1.6×10^{-13}	1.2×10^{-13}	3.8×10^{-17}	4.5×10^{-20}	1.9×10^{-24}	1.8×10^{-27}	1.8×10^{-30}
1 amu*	=	9.3×10^{8}	930	1	1.5×10^{-3}	1.5×10^{-10}	1.1×10^{-10}	3.6×10^{-14}	4.1×10^{-17}	1.7×10^{-21}	1.7×10^{-24}	1.7×10^{-27}
1 erg	=	6.2×10^{11}	6.2×10^{5}	670	1	10^{-7}	7.4×10^{-8}	2.4×10^{-11}	2.8×10^{-14}	1.2×10^{-18}	1.1×10^{-21}	1.1×10^{-24}
1 joule	=	6.2×10^{18}	6.2×10^{12}	6.7×10^{9}	10^{7}	1	0.74	2.4×10^{-4}	2.8×10^{-7}	1.2×10^{-11}	1.1×10^{-14}	1.1×10^{-17}
1 ft.lb	=	8.4×10^{18}	8.4×10^{12}	9.1×10^{9}	1.4×10^{7}	1.4	1	3.2×10^{-4}	3.8×10^{-7}	1.6×10^{-11}	1.5×10^{-14}	1.5×10^{-17}
1 Cal†	=	2.6×10^{22}	2.6×10^{16}	2.8×10^{13}	4.2×10^{10}	4.2×10^{3}	3.1×10^{3}	1	1.2×10^{-3}	4.8×10^{-8}	4.7×10^{-11}	4.7×10^{-14}
1 kWh	=	2.2×10^{25}	2.2×10^{19}	2.4×10^{16}	3.6×10^{13}	3.6×10^{6}	2.7×10^{6}	8.6×10^{2}	1	4.2×10^{-5}	4.0×10^{-8}	4.0×10^{-11}
1 MW day	=	5.4×10^{29}	5.4×10^{23}	5.8×10^{20}	8.6×10^{17}	8.6×10^{10}	6.4×10^{10}	2.0×10^{7}	2.4×10^{4}	1	9.6×10^{-4}	9.6×10^{-7}
1 g	=	5.6×10^{32}	5.6×10^{26}	6.0×10^{23}	9.0×10^{20}	9.0×10^{13}	6.6×10^{13}	2.1×10^{10}	2.5×10^{7}	1.0×10^{3}	1	10^{-3}
1 kg	=	5.6×10^{35}	5.6×10^{29}	6.0×10^{26}	9.0×10^{23}	9.0×10^{16}	6.6×10^{16}	2.1×10^{13}	2.5×10^{10}	1.0×10^{6}	10^{3}	1

1 Btu (British thermal unit) = 1055 J $\approx 10^{3}$ J.
1 HP.hr = 2.7×10^{6} J = 0.746 kWh (1 HP = 746 W).
* 1 amu (Atomic mass unit) = 1 g divided by Avogadro's number.
† The Calorie, not the small calorie.

TABLE 2
Examples of Energy Conversion Processes

To / From	KINETIC	GRAVITA-TIONAL	ELASTIC	THERMAL	ELECTRIC & MAGNETIC	RADIATIVE	CHEMICAL	NUCLEAR
Kinetic	Elastic collisions Musical instruments Windmills	Motion under gravity (e.g. pendulum)	Spring bumpers	Friction (e.g. brakes on car or bicycle)	Electric generator Telephone	X-ray production		
Gravitational	Falling objects Tides			Contraction of a star	Hydroelectric generator	Gravitational waves (?)		Formation of neutron star
Elastic	Spring clock Catapult or bow & arrow	Trampoline or springboard		Compression of a gas	Piezoelectricity			
Thermal	Heat engines Wind Ocean currents	Convection	Bimetallic strip		Thermo-electricity	Radiation from a hot body Incandescent light bulb	Endothermic reactions	
Electric & Magnetic	Electric motor Loudspeaker Thunder Particle accelerators	Electric lift (elevator)	Electro-striction Magneto-striction	Immersion heater	Transformer	Lightning flash Radio & TV Fluorescent light LEDs	Electrolysis Electroplating Charging a battery	
Radiative	Photoelectron emission			Absorption at surface Microwave oven	Solar cell	Fluorescence	Photosynthesis Taking a photograph	
Chemical	Rocket engine Muscle action			Exothermic reactions	Storage battery Fuel cells	Fireflies		
Nuclear	Spontaneous fission			Nuclear fusion Nuclear power plants		Gamma-ray emission Stars		

3

Teaching Methodology in Energy Education

D. R. BALURAGI

LVD College, Raichur, India

The pedagogic method of introducing the concept of energy in schools is generally restricted to the definition of energy with a few forms and their interrelationships. Electricity via commonly available dry cells, and animal energy through examples of agricultural practice in rural areas, are two of the suggested starting points. Water lifting devices and photosynthetic conversion of solar energy to food materials in plants are also common ideas to start with. Commonly available spring toys may be used to identify potential energy. Moving balls will do to illustrate kinetic energy.

Wherever possible, students should be taken on a field trip and examples of energy utilisation should be shown to them. In the class room itself, some of the concepts could be made clear by bringing demonstration objects such as a carom board, play bow and arrow, magnets, etc.

A useful exercise for the teacher is to ask the students to narrate their day's activities before coming to school and analyse each in terms of energy utilisation. The Sun is the source of all our energy in this world. Experiments could be designed or demonstrations in the field shown to students, so as to highlight the importance of solar energy. Less costly solar water heaters and solar cookers may be constructed by students with the assistance of the teachers. Alternatively, students may be taken to a domestic or industrial site where solar panels are used.

It would be a useful programme to ask the students to fill out a questionnaire wherein they indicate the energy used for cooking in their houses. Concepts of efficiency of cooking fuels may be explained and, if possible, illustrated. The difference in practice between rural and urban areas may be made clear. Pictures and charts on the following areas may be used to illustrate the applications and varieties of energy:

— Energy food or fuel foods.

— Energy for transportation, rural and urban.
— Energy in houses—fuel efficient stoves, etc.
— Energy for agricultural operations such as sowing, ploughing, transplantation, harvesting, drying, etc.
— Water lifting using bullock drawn devices, electricity, diesel and/or biogas generators, solar pumping, etc.

Objectives

The objectives of energy education for students are as follows:

— Becoming acquainted with various forms of energy and their interconversion.
— Learning about the role of energy in their daily lives.
— Acquiring knowledge about commercial and non-commercial energy and the difference in rural and urban practices.
— Learning about energy crisis and methods of overcoming it.
— Becoming aware that energy is not infinitely available—through this the methods of conserving energy, augmenting it.
— Identifying energy options for future.

Modes of Implementation

Modes to aid student understanding include:

1. Field trips.
2. Intensive study of energy utilisation in a house (rural/urban).
3. Energy education kits and packages to teach academic requirements. (Karnataka Rajya Vijnana Parishat is producing one such package.)
4. Community of students may build a small solar water heater for their hostel/school use.
5. Groups of students may be deputed to measure wind speeds at different places using anemometer.
6. Audiovisual aids such as films, slides, transparencies, charts, models.

4

Energy Curriculum Development Guidelines

D. F. KIRWAN

University of Rhode Island, USA

Here are the basic steps involved in energy curriculum development, with appropriate strategies to assist energy educators to overcome potential problems and achieve their objectives:

Identify curriculum goals.
Assess the students' knowledge.
Consider the students' motivation to learn.
Define and organise the content.
Select materials.
Design and/or select appropriate strategies.
Assess the students' learning.

The curriculum should reflect the intent, capabilities, resources, and interests of the institution where the programme will take place, and of the teachers who will implement the curriculum and/or play a role in its development. Many teachers may have difficulty teaching energy topics due to their lack of conceptual understanding and/or involvement.

Identifying the Curriculum Goals

Energy educators or school energy curriculum design teams should pay particular attention to the important activity of designing curriculum goals. The curriculum goal must answer the question; what do students need to learn and/or be able to do? For example, what would students need to know in order to: (1) build a solar cooker, (2) understand the basic principles of energy conservation, (3) assist their families to insulate their homes? The detailed analysis of these questions would provide much useful information for curriculum development.

Energy educators should also address several goal-related questions early in the planning phase.

What Should be the Scope of the Curriculum?

How comprehensive a view of energy education need the curriculum offer? For example, is it necessary to educate students in energy conservation in addition to presenting local, national, and international aspects of energy supply and demand? Curriculum planners should pay considerable attention to the scope of learning activities that the programme will offer to students. The scope should not be so broad that it cannot be carried out effectively. Similarly, it may not be necessary to focus equally on such topics as energy conservation and renewable energy development and applications.

It is suggested that teachers without significant experience in energy education take a conservative approach to curriculum goals and scope. Do not attempt too much. Remember that there is usually a tendency to overestimate programme capabilities and resources. It would be a better idea to move into this new area of activity gradually and only increase the scope of educational activities when experience has proved the feasibility of doing so.

What is the Organisational Capability?

Curriculum goal-setting should not be accomplished without a careful assessment of organisational capability. Schools need to assess their present capability to begin energy education programmes, based on their recent experiences in energy, educational, and community development programming. What could schools reasonably accomplish without straining their capabilities?

What Resources are Available?

Similarly, having tentatively established programme goals, energy educators should then determine the resources that are needed, their costs, and their availability. This is a useful planning step even for the smallest programme, and one which indicates whether there are sufficient resources available to do the specific programme. Educators should try to be practical and inventive as they do this planning!

Teachers can research this important question by developing a list of programme-supporting activities that are specifically oriented to the curriculum goals and identifying the resources that are required for each. For example, a programme that intends to educate students to build a windmill will need materials, skilled teachers from the school and/or other instructors, sufficient space in the school or at another facility, and amounts of money sufficient to cover these needs and other miscellaneous requirements.

Depending on the goals of the programme and the scope of the curriculum, the amount of required resources could be considerable.

How Much Time is Needed?

What is the duration of the programme? How much time will school teachers have to work on a particular energy education programme? How much learning time is available to students, given class, homework, and outside responsibilities? How much learning can students, their families, and selected community groups be expected to accomplish?

How Knowledgeable are the Instructors/Teachers?

Energy education programme planners also need to assess the knowledge and expertise of school teachers. Are they familiar enough with the material to be taught, or will they themselves need preliminary training? If so, what types of training for what period of time? Where can this training be obtained, and at what cost? Judged against the constraints, is the training of teachers a feasible activity? If it is not, planners may have to modify programme objectives to accommodate what is realistic.

What Considerations Should be Given to In-service Teacher Training?

Energy education programmes that need in-service teacher training should select a particular training programme or institution carefully, for in-service teacher training programmes in the energy field cover a wide range of subjects and teaching methodologies. Not all of these may be appropriate to the more practically oriented needs of those energy education programmes in which instructors need to be involved in building and installing energy hardware.

Existing faculty development programmes in energy can provide an introduction to energy as a broad issue. These programmes should provide a direct link between what is perceived as an energy problem and the local situation. Appropriate programmes will include more practical activities.

Various governmental, industrial, and other agencies could provide funding for the establishment of faculty development programmes. These programmes could organise and conduct workshops for teachers which might teach them basic energy skills, engage them in more advanced energy training, get ideas for curriculum approaches from them, and involve them in conferences. These workshops could be held in remote areas or in other areas which have been identified as locations where the need is greatest.

The questions above are samples of typical goal-related considerations which must be addressed in the planning stages. The curriculum planners may need to address other questions which are less generic in nature and may depend upon the demographic and geographic characteristics of their region.

Assessing the Students' Knowledge

As the programme begins, instructors, in order to establish realistic learning goals for the programme's duration, may want to discover what the students already know about energy. Students' knowledge should be assessed from the perspective of the physical sciences. Their mathematical knowledge should be determined, and basic understanding of energy conservation and renewable resource development and applications explored. While some students will know more than others, instructors will surely be able to determine a general level of energy-related knowledge and be able to begin to pinpoint the strengths and weaknesses of the students.

Assessing the Students' Motivation to Learn

At the same time, energy education programme instructors will also want to determine the students' interest in the programme. Energy educators have reported that student motivation is one of the basic problems they encounter. In general, formal education is frequently not directly related to students' interests nor to their day-to-day experience in their communities. What is true of formal education may also be applied to energy education programmes. It is not difficult to understand how, for example, those programmes dealing with global energy issues are not particularly relevant to students whose knowledge of geography and world politics may be sketchy at best. This is not to say that awareness of the world energy context is not important, but that it must be communicated in an appropriate way to the students. For example, if teachers use a case study approach centring around the practical problem of paying energy costs, the very immediacy of that local problem could serve as a stimulus to their students to better understand the internationally-based causes for oil prices and other energy costs. Inappropriate curricula and teaching that leave students passive in the classroom, which are difficult to understand, or which lead to no relevant goals in the students' estimation, serve to reinforce a negative perception on the part of students. Attacking this motivation problem involves changing the negative perception to a positive one.

The students' interest could be developed through a programme which, drawing on multi-disciplinary approaches, is at once personal, practical, and purposeful. Making the programme a personal experience involves structuring it around the perceptions and experiences of the students. Relating energy to the individuals, their home and community, through a case study approach, for example, is one way to do that.

Secondly, energy educators and instructors can emphasise the development of a personal yet authoritative teacher–student relationship. A personal approach encourages students to have a positive self-image, which continuing activity and the completion of projects can serve to reinforce. Continued reinforcement by the instructors is perhaps the most important component essential to motivating students.

Making the programme practical means having students build things. Practical activities give learners a sense of control over their own lives. Practical activities also provide learners with the knowledge that they can produce something useful, with the resulting feeling of self-esteem because they can produce objects of value. The more practical the projects the students complete, the stronger the students' positive feelings should become.

Closely related to this aspect of practicality is the purposefulness of the students' activity. Not only is this work useful to the students' own lives, potentially conferring the possibility of career development, but it is also useful to their families, friends, neighbours, and community. Schools should therefore strongly consider the active involvement of students in the programmes. Such involvement may have a number of dimensions, including programme planning, design and re-design of activities, construction, model building, demonstrations, and performing home energy related activities with the participation of their families. It could also include participating in and organising various community outreach and public information activities, including poster contests and school, street, and community fairs.

Effective teaching by instructors in the programme will also serve to motivate students. Teachers need to demonstrate their practical command over the energy-related material to be taught. Teachers who are not qualified to carry out the various aspects of the programme, whether academic, manual, or a combination of the two, will soon find themselves at a decided disadvantage with the students, who are usually quick to note such deficiencies. Instructors also need to create and promote an atmosphere of support for students. While this is not the only or the most important role for teachers, it is nevertheless an essential one for programmes whose basic strategy it is to involve the students.

Defining and Organising the Content

The instructors are now ready to define and organise the specific content of the course of instruction. These specifics should follow logically from the curriculum goal. The construction of an objective vs. time graph indicating what the students should know and/or do by a certain date will aid in this process. While the content determines the bulk of the information students need to assimilate in order to reach a particular objective (intermediate goal), activities should be selected which foster content mastery. In other words, activities should be incorporated into the instructional design based upon the extent to which a particular activity will contribute to achieving the educational goal.

As an example, suppose an instructor established as an intermediate curriculum goal: "At the end of a four-month period, students will be able to design and construct working models which demonstrate the conversion of solar energy to thermal energy." In order to achieve this objective, the instructor decided the students would need to study and explore the following course content:

Principles of heat loss and gain.
Principles of ventilation.
Fundamentals of insulation.
Heat storage systems.
Qualities of materials.
Physical design principles.
Elementary carpentry skills.

The instructor then selected numerous activities specifically related to enhancing the probability that students would master each of the identified content areas.

It might be useful at this point to state that it is not effective to take an overly strict approach to the implementation of the curriculum. Energy educators should design a certain amount of flexibility into the curriculum, with regard to both goals and content. Students and teachers can then take advantage of unanticipated opportunities that might enhance the programme and can also improvise and redesign aspects of the programme that may need modification. There are many other situations in which too strict adherence to the orginal curriculum would have more negative than positive effects on the outcome of some aspect of the programme. It is probably best for energy educators to design a moderately flexible curriculum which can be modified on the basis of careful monitoring of the programme's progress.

Selecting Materials

Energy education programme instructors need to select the materials —texts, audiovisuals, demonstration models—that will best illustrate the content that has been defined and organised in the preceding step. The materials will thus reflect that step and directly assist the students to achieve their educational goals.

The materials that instructors choose will most likely be multi-disciplinary in nature, derived from the many ways in which energy influences our lives, as reflected in the disciplines of natural science, mathematics, social science, art, and philosophy. The multi-disciplinary approach is useful because it allows the school to do energy studies and activities within the general confines of the established curriculum without imposing a significant burden on the teachers. This approach encourages teachers to both incorporate energy-related information into their own classroom activities and to contribute their specialised knowledge and skill to potentially all school energy-related events. For example, an art teacher would be instrumental in the development of a school energy poster contest, or a history teacher might cover the historical aspects of energy use. Schools decide the nature of this multi-disciplinary effort themselves, based on a common discussion of possible programmes, available personnel, and appropriate resources.

Designing and/or Selecting Appropriate Strategies

The development of appropriate and effective strategies is one of the most important activities for programme planners and educators. Some of these strategies are as follows:

Do Hand-on Activities

This is a highly recommended approach which motivates students and enables them to better understand the course content. Energy education planners would do well to recall the old adage: "Tell me .... I forget; Show me .... I remember; Involve me .... I understand!" One often overlooked rationale for the manual activities strategy in the classroom is that it teaches students to work cooperatively on team projects, encouraging the more skilled to teach and assist the less skilled. Additionally, hands-on activities permit students to take a more active and independent role in their own learning. Hands-on activities also promote stronger self-images through the completion of short-term, tangible projects. Outside the classroom, the active involvement of the students in practical, energy-related work and/or learning exercises and activities might take the form of energy poster contests, project or activity demonstrations throughout their community to other schools, and interested groups and at local and regional gatherings.

When energy education planners opt for this strategy, they need to carefully consider those programme resources which might be necessary to successfully involve students in practical activities.

First, to implement a hands-on approach the programme needs instructors who can perform the specific manual and mechanical activities without hesitation and who also know the scientific principles that underlie those activities. If the intended teachers do not have this background, it might be useful for the programme to arrange for suitable training. If such training is not feasible, programme planners might consider implementing a team system, whereby skilled practitioners are available to support the more inexperienced instructors when the latter need assistance. In addition, the participation of local people who have done projects on their own as backup instructors and/or guest speakers is also desirable. These individuals lend credibility to the new programme because of their prior successes.

In addition, the instructors should be capable of breaking down the activity into its component steps and logically presenting these to the students. This ability may require some practice on the part of teachers for whom this type of instruction is relatively new.

Second, such activities will require materials that are durable enough to allow many students to repeat the exercises. Programmes should obtain sufficient materials to permit a number of students to perform specified activities in order to develop a productive momentum in the class.

Third, in order to improve the effectiveness of instruction, it might be a good idea to introduce specific manual activities with an audiovisual presentation that shows others working on various stages of a specific project, including views of the finished product.

Conduct Workshops

A workshop is a brief educational programme for small groups which emphasises group participation in energy-related activities. This approach has numerous advantages. Educators, drawing on their previous energy-related experiences could design workshops to be presented in classes at either primary or secondary levels. Once educators have given a small number of such workshops, they will begin to develop a basic workshop model within the energy field which will be applicable to a broad range of target audiences. Workshops, which take relatively less time to prepare than classroom curricula and lesson plans, often have more of an impact than the more routine format of classroom instruction.

Make Site Visits

For low-income urban and rural students, site visits represent opportunities to increase their knowledge of energy-related matters, the energy options available locally and regionally, and the energy problems of the communities in which they live. Energy education programmes indicate that site visits are well appreciated by students. In general, there is a broad range of potential site visits for students to make.

Students will learn more from these site visits if the teachers make sure that the site visit takes place within an overall learning context. Prior to the visit, the students should have some general understanding of the purpose of the site visit and its relationship to the energy material being studied in class. Teachers can convey this understanding to students through preparatory class discussions. Similarly, a post-site visit activity such as a class discussion, a small group debate, or individually written reports will help students to reinforce the impressions gained at the site visit, gain answers to new questions, and put the experience into perspective.

Broadly considered, visits offer many opportunities to energy education programmes. In terms of numbers of students participating, an entire class, groups of students, or individual students may make site visits. One objective of a site visit might be to interview energy-related workers in the community to learn more about specific energy problems. Such visits could lead to the development of individual or group work/study projects.

Use Audiovisual Materials

The use of audiovisual materials, to accompany lectures and as an integral part of workshops, is also an effective method of communicating energy information to students. Many energy educators show slides and films as part of their presentations to schools and communities. They also make use of other types of audiovisual materials, including working demonstration models of energy devices. More advanced students could make energy presentations to various citizens' groups. In fact, some programmes have their students actually build models for use in family and community outreach energy education activities.

Assessing the Students' Learning

Assessing the students' learning is the last major step to be carried out by energy programmes. This assessment activity, which is also called evaluation, takes place both during and at the end of the programme, and generally consists of the design, writing, and administering of a variety of measures to assess the learning of energy-related knowledge, skills, and attitudes.

Evaluation serves a number of useful purposes. It indicates the quantity of the students' learning and provides a firm rationale for changes in programme objectives and/or instructional activities. It allows the students to record their own progress, thereby increasing their motivation. Evaluation also provides a feedback loop which serves as a stimulus and may provide the impetus for instructors to improve future efforts. Evaluations can serve to provide documentation of the programme's teaching/learning experience so that other programmes may benefit from it.

School curriculum planners/instructors should begin to think about evaluation as early as the design phase of the programme. *What, when,* and *how* to assess the students' learning are perhaps the major questions for educators to consider. Given the great variety of energy education programme objectives, approaches, and activities, it is not possible here to make more than a few suggestions in response to the following questions.

What is to be Assessed?

Assessment should be primarily concerned with the degree to which students are achieving the programme objectives, i.e. the intermediate (middle-term) and specific (short-term) curriculum goals. If curriculum planners/instructors have broken these goals into specific learning units, they should be able to assess the students' progress without too much difficulty.

When Should such Assessment take Place?

Assessment of student learning by instructors should take place at regular intervals to provide systematic and continuous flow of information back to the students and frequently enough to give both students and instructors ample time to make any necessary modifications in their learning and teaching activities.

How Should Assessment be Done?

There are a variety of test and non-test methods that schools can employ to assess their students' learning. Because of the uniqueness of many locally developed programmes, there will be very few appropriate standardised tests for energy education programme instructors to use. Instructors will most likely have to design their own tests, which reflect the programme's specific learning goals, and the particular strategies emphasised.

Most programmes will emphasise the teaching of energy-related knowledge and skills through the performance of specific energy-related tasks and projects. Assessing the students' achievement of these tasks may call for different types of measures. A pen-and-paper test may not be an appropriate one. Instructors will have to develop tests based on criteria that are appropriate to the specific performance of a practical, skill-oriented task. Here are some criteria that instructors may wish to apply:

- Was the time the student spent on task commensurate with the time required to satisfactorily complete the task?
- Was the effort expended by the student on the task commensurate with the work required to satisfactorily complete the task?
- Was there evidence that the student understood the specified energy and other task-related concepts?
- Was there evidence that the student made appropriate plans to complete the task satisfactorily?
- Was there evidence that the student completed the task using creative or innovative methods?
- Was there evidence that the student was effective in communicating task development and results to the class and/or instructor?

SECTION B

Energy Education at the Primary Level

Introduction

This section and the following section contain specific and detailed activities of interest to the teachers of energy. This section focuses on primary activities. An inquiry approach is taken and each activity develops or reinforces the process skills which are appropriate for this age group. The best introduction to this section, however, is contained in the first article which the leader of the primary team has written which addresses energy education and learning at the primary level.

5

Energy Concepts in Primary Education

I. WINTER

Claremont High School, Tasmania, Australia

The need to address ourselves to this topic is based on the following self-evident premises:

All societies are users of energy.
Awareness of these societies of the role of energy in their lifestyles is crucial.
The ability to distinguish between renewable and non-renewable energy sources is a basic need.
The ability to differentiate energy usage on the basis of cost efficiency and function efficiency will determine rates of progress of a developing society.
Effective education will develop the previous three attributes.
Effective education will lead to innovation and invention.
Education in many societies does not extend beyond the primary level.

Just as science topics are generally avoided by primary teachers, energy awareness at least in formal terms is not a well developed pedagogical art. Many Western countries have sophisticated primary energy instruction materials but these are sparingly used and then only by well-versed teachers. A leading British observer has stated that primary teachers should avoid teaching energy topics as the concept is felt to be too abstract for the majority of primary students who are still operating at the concrete stage of Piagetian development. In the light of the above listed premises, this last argument needs to be considered and rejected. Appropriate classroom action plans and student materials will support this rejection.

To produce an action plan, the following questions should be answered:

1. Can energy be defined in simple ways such that all elementary teachers will easily identify with it?
2. Can the concept of energy be rationalised to encompass only everyday, concrete experiences?

3. Can "hands on" activities be devised which incorporate only the simplest materials, to demonstrate energy concepts?
4. Can all the appropriate energy conversions which will complement our concept of energy and stimulate primary students towards attainment of "energy literacy", be identified?

If each of these questions cannot be answered with convincing affirmation, the effort falters right here. However, many educators will have further and more emphatic evidence. The following solutions to the above questions are offered here.

1. Let energy be defined as simply a "change-maker". A plant grows, a candle burns, a wheel turns, water evaporates: all changes, simple and complex, have an energy association. Perhaps at elementary level the word "energy" need not be introduced until late in the programme. "Change-maker" removes the abstract at least in name.
2. The examples mentioned above are indeed everyday concrete experiences. It should not be difficult to extend this list and from it draw those best suited to the needs of specific countries or specific localities.
3. Consider a rubber band being stretched and relaxed very rapidly. Changes will occur and can be observed occurring. Sound and heat will be generated.

The ideas which follow were generated and discussed by the primary science participants in the Energy Group at the Bangalore Conference. Most of these activities describe simple, readily available materials and should be seen as starting points for further investigation and discussion in the classroom.

4. We identify the kinds of energy appropriate to primary level as being:

heat	movement
sound	electrical
light	magnetic

In some countries the term "atomic" or "nuclear" energy may need to be simply treated if local or national issues demand it.

Students will be quick to recognise change-making items which convert one kind of energy to another:

fire = chemical energy to heat energy and light energy,
lamp = electrical energy to heat energy and light energy,
drum = movement energy to sound energy,
etc.

With these encouraging responses to the earlier questions, it is necessary to establish an implementation strategy. It should be accepted that teaching energy in the primary school is necessary for developing basic energy awareness, energy literacy, and energy conservation.

To achieve this, it should be treated as any other topic in the primary science field. An inquiry approach should aim to develop:

Attitudes	Highest priority
Skills	High priority
Knowledge	Least high priority

The presentation of the curriculum should encourage:

(a) *Positive attitudes*: Students want to participate.
(b) *Learning how to learn*: Students acquire understanding and use of process skills.
(c) *"Hands on" activities*:

I hear	=	I forget
I see	=	I remember
I do	=	I understand

(d) *The need for scientific literacy is universal*: Boys and girls should participate equally.

Teachers should be familiar with the following process skills and structure lesson presentations so that students learn to:

Observe	Control variables
Infer	Experiment
Predict	Interpret data
Classify	Hypothesise
Measure	
Communicate	

The first six of these skills (first column) are basic process skills and can be developed from the earliest years of education. The last four of these skills are integrated process skills, i.e. combinations of the first six may be more relevant for higher primary grades.

The activities presented on the following pages have been developed with the inquiry strategy in mind. A situation is presented which is designed to stimulate questions from the students which can be answered by the students undertaking further investigation. Each activity involves the use of one or more of the above process skills.

6

Energy Activities in Primary Education

D. R. BALURAGI, H. HARNAES, J. A. RODRIGUEZ, R. SUPORNPAIBUL and I. WINTER

The following are a series of activities which have been found effective at the primary level. Throughout them the teacher should constantly reinforce the concept of energy as "change-maker".

ACTIVITY TO STIMULATE INQUIRY INTO MAGNETIC ENERGY

I. Winter
Claremont High School, Tasmania, Australia

Objective: Students should be able to infer that something which cannot be seen can exist.

Materials: Ordinary paper clip, cotton thread, magnets, and other magnetic and non-magnetic materials.

Initial setup: One end of a thread is attached to a paper clip and the other is taped onto a table top. A magnet is fixed to a second table close enough so that when the paper clip is brought near the magnet, it will be suspended in air with a noticeable gap between the clip and the magnet.

Activity:

1. Ask students to construct at least six questions all starting with "What would happen if". For example:
   — What would happen if the cotton thread is cut?
   — What would happen if another magnet was brought near the paper clip?
   — What would happen if we use string instead of thread?
   — What would happen if a piece of (copper, glass, paper, steel, aluminium, etc.) is placed between the magnet and the paper clip?
2. Having asked each question, students should make a prediction about what they think might happen.

3. Students should then set up their own magnet–paper clip arrangement and proceed to test their predictions by experimenting. For example, students can place different materials in the gap and determine which materials will make the paper clip fall.

One obvious outcome of this activity is the identification of materials which are affected by magnetic energy.

ACTIVITY TO STIMULATE INQUIRY INTO HEAT ENERGY

R. Supornpaibul
Chulalongkorn University Demonstration School, Bangkok, Thailand

Objective: After completing this activity students should recognise that unseen energy exists.

Materials: Heat source (hot brick or charcoal stove), glasses, water, thermometers.

Activity:

1. Place the heat source on the floor.
2. Place two or three identical glasses of water different distances from the heat source.
3. Have the students compare the water temperature in each glass by use of a thermometer.
4. Predict what temperature might be found at some intermediate distance from the stove.
5. Test the prediction by measuring the temperature of a glass of water at that distance.
6. Repeat the experiment with the glasses at the same distance from the heat source, but arranged along different radii outward from the source.

ACTIVITY TO STIMULATE INQUIRY INTO NUCLEAR ENERGY

I. Winter
Claremont High School, Tasmania, Australia

Note: This activity should be demonstrated by a secondary teacher or an education officer attached to a nuclear energy authority. Appropriate safety aspects should be described, illustrated, and discussed.

Objective: The students will observe that some substances have a special "change-making" quality.

Materials: Geiger counter, radioactive sources, paper, lead, wood, and other materials.

Activity:

1. A Geiger counter is placed at positions 10 cm, 20 cm, and 50 cm away from the source. Students will hear the sound and recognise that:
   (a) the sound is heard only when near the source,
   (b) the count of sound is diminished with distance.
2. Keeping the counter at 10 cm, various materials can be placed between it and the source. Students will recognise that some substances will "protect" the counter while others will not.

ACTIVITY TO STIMULATE INQUIRY INTO MOVEMENT ENERGY

D. R. Baluragi
LVD College, Raichur, India

Objective: The students will infer that some "change-making" action implies the existence of energy.

Materials: Eye dropper, ink, water and other fluids, several sheets of non-absorbent paper, pinwheel.

Activity:

1. Place the piece of non-absorbent paper on a table.
2. Using the eye dropper, allow a single drop of ink to fall on the paper from a height of 20 cm.
3. Using the eye dropper, allow a single drop of ink to fall at a different spot on the paper from a height of 40 cm.
4. Using the eye dropper, allow a single drop of ink to fall at a different spot on the paper from a height of 60 cm.
5. Do not move the paper, let the ink dry.
6. When the ink is dry, measure the width of the impact spot and count the number of "spines" around each spot.
7. Record all the observations and discuss the inferences.
8. Test the inferences by repeating the experiment.
9. Have the students predict what size the impact spot might be at 80 cm, 100 cm, etc.
10. Promote further questions such as:
    (a) What would happen if we used a "thicker" liquid such as paint?
    (b) What would happen to the speed of a pinwheel if drops of water are dropped on it from different heights?
11. Construct a water wheel to run on a tap. Perhaps this can be connected to a small generator (from an old-style telephone) to light an electric bulb.

Students should be able to infer that movement of matter can cause changes and that energy is associated with moving objects.

ACTIVITY TO STIMULATE INQUIRY INTO HEAT ENERGY FROM THE SUN

H. Harnaes
University of Oslo, Norway

Objective: Students should be able to infer that the sun produces heat energy.

Equipment: One small-necked bottle, a balloon, a pan of hot water, a pan of cold water, sunshine.

Activity:

1. Fix a deflated balloon over the mouth of a bottle.
2. Place the bottle in a pan of hot water.
3. Have the students describe what happens.
4. Ask the students:
   — What would happen if you placed the bottle in a pan of cold water?
   — What would happen if you placed the bottle back into the pan of hot water?
   — What would happen if the bottle were placed in the sunshine instead of the hot water?
5. When students have made their prediction, they should test it.

ACTIVITY TO STIMULATE INQUIRY INTO HEAT ENERGY

R. Supornpaibul
Chulalongkorn University Demonstration School, Bangkok, Thailand

Note: Be sure to exercise proper safety techniques when boiling the liquids.

Objective: Students will investigate some of the properties of liquids when heated to the boiling point.

Materials: Transparent flask suitable for boiling liquids, heat source, water, vegetable and animal oil.

Activity:

1. Fill the flask with water and place on the heat source.
2. Have the students observe and predict what is going to happen:
   — What happens inside the flask?
   — How long does it take before the water starts to boil?
   — If the students put some more water into the flask, will it affect the time?
3. Have the students measure the boiling point of water.
4. Several investigations should start at this point:
   — If the students put some more water into the flask, will it affect the boiling point?
   — Are the boiling points of plant oil and animal oil the same?
   — Which one of these will take longer—to boil water or to boil oil?

ACTIVITY TO STIMULATE INQUIRY INTO SOLAR ENERGY

I. Winter
Claremont High School, Tasmania, Australia

Objective: Students will investigate ways of trapping the heat energy delivered by the Sun.

Materials: Two identical bottles, paint, thermometer, water and sunshine.

Activity:

1. Two identical bottles are painted, one black and one white. Each is filled with the same volume of water and placed in a clear, sunny position. After a period of time, e.g. 1 hour, a thermometer is placed in each bottle.
2. Students should observe that the temperature of the black bottle is higher than that of the white bottle. Several investigations should start at this point.
   — What would happen if the bottles were painted on the inside instead of the outside?
   — What would happen if the bottles were clear glass, but the water was coloured black in one and white in the other?
   — What would happen if one bottle was wrapped in foil?
   — What would happen if each bottle had a piece of foil taped to the back half only of the bottle?
   — Conduct a competition to see who can make the best solar "heat trapper" using identical bottles, identical volumes of liquid, and any other materials they may require.

ACTIVITY TO STIMULATE INQUIRY INTO SAVING ENERGY

I. Winter
Claremont High School, Tasmania, Australia

Objective: Students will design their own experiments using the concept of conserving heat energy.

Materials: Cups of various shapes, water, milk, thermometers.

Activity:

1. Students will be asked to determine what is the best shape for a coffee cup in order to retain the heat for the longest time?
2. Students collect cups of various shapes and pour equal volumes of water at the same temperature.
3. The temperature is measured at regular intervals (e.g. 3 minutes) and recorded.
4. Students make a decision on the best shape. This should lead to the following kinds of investigations:
   — Which material is best for a coffee cup: metal, china or glass?
   — How important is thickness of the cup wall?

— Does it help if the cup is rinsed with hot water before making the coffee?
— What difference does it make if milk is added before or after mixing the coffee?
— What can we do to the cup to reduce heat loss? (Wrap the cup in various materials: paper, foil, cloth, etc.)

ACTIVITY TO STIMULATE INQUIRY INTO ELECTRICAL ENERGY

I. Winter
Claremont High School, Tasmania, Australia

Objective: Students will demonstrate the meaning of a complete circuit.

Materials: Cards or poster paper, metal foil strips, scissors, battery, wires and light.

The teacher will need to make up several cards using the metal foil. The metal foil is laid in strips on the card in various disconnected patterns. Cut several holes (about six in a symmetric pattern) in a second similar sized card. A little planning on the teacher's part when laying the metal tape will allow the metal foil to show in the holes when the second card is placed on top of the first. The edges are then taped. This is now a "mystery card".

Activity:

1. Students use a battery, wires and light bulb to discover the arrangement of the foil inside the card.
2. Students can make their own mystery cards of appropriate difficulty.
3. Students can use this idea to make an "electric quiz".

ACTIVITY TO STIMULATE INQUIRY INTO MOVEMENT ENERGY

J. A. Rodriguez
Caracas, Venezuela

Objective: Students will observe that the further an object falls, the more energy it possesses.

Materials: Two blocks of wood, a board wide enough for the blocks to slide without falling off the sides.

One block is positioned at an end of the board on a table. The board is then tilted up to an angle of about thirty degrees and clamped at that angle, making an "inclined plane".

Activity:

1. A block of wood (A) is positioned on the board about 50 cm above the tabletop and released from that position so that it slides down and hits the other block (B) which is on the tabletop.

2. Students measure how far B travels after being struck.
3. The experiment is repeated, this time releasing the block from positions higher and higher above the tabletop.
4. Some examples for further investigation are:
   — What happens to the energy of the sliding block if the surface is made smoother? rougher?
   — What happens if block A is larger than block B?
   — What happens if block A is smaller than block B?
   — Repeat the experiment using rolling spheres (marbles on "tracks" on corrugated cardboard is a good idea).

ACTIVITY TO STIMULATE INQUIRY INTO STORED ENERGY

R. Supornpaibul
Chulalongkorn University Demonstration School, Bangkok, Thailand

Objective: Students will recognise that an object has more "change-making" effect the higher it is.

Materials: Two jars or pails for water, plastic tube for siphon.

Activity:
1. A simple siphon is constructed between a fixed volume of water is placed in a jar about 40 cm above a second jar and allowed to run into the second jar.
2. The time taken is measured. The following may also be investigated:
   — What happens if the height difference is 40 cm? Also try 60 cm.
   — Use these results to predict the time for 80 cm, 100 cm, 50 cm.
   — Does it make any difference if we use another liquid, e.g. oil?
   — Instead of using a plastic tube, use a strip of absorbent paper. Compare the starting amount and the finishing amount.

ACTIVITY TO STIMUALTE INQUIRY INTO SOUND ENERGY

H. Harnaes
University of Oslo, Norway

Objective: Students will discover that different materials are affected in different ways by sound energy.

Materials: Two empty tin cans, string, wire, paper cups.

Activity:
1. Two empty tin cans are connected by a piece of string. This is done by punching a hole in the bottoms of the cans and knotting the string with a knot on the inside of the cans so that the string does not pull through when pulling the cans apart until the string is stretched taut.

2. If a student speaks into one, he/she can be clearly heard by a student holding the other can to his/her ear.
3. Investigations to be carried out could include:
   — What happens if we use a longer thread?
   — What happens if we use a thicker thread?
   — What happens if we use a metal wire?
   — What happens if we use paper cups instead of tin cans?
   — What happens if we change the size and shape of the tins?
   — Who can make the best "telephone"?

ACTIVITY TO STIMULATE INQUIRY INTO SOLAR ENERGY

J. A. Rodriguez
Caracas, Venezuela

Note: This activity results in a small fire to a piece of paper. The teacher should exercise the proper amount of caution when doing this activity.

Objective: Students discover that light from the Sun and heat from the Sun have powerful "change-making" properties.

Materials: Converging lens, paper, cloth, glass, sunshine.

Activity:

1. Place a cover lens in front of a piece of paper such that a sharp point of light is obtained and the paper catches fire.
2. Students should now investigate:
   — How far from the lens must the paper be?
   — What happens if we use cloth instead of paper?
   — What happens if we try to heat a small container of water this way?
   — What happens to the paper if we place ordinary glass in front of the lens?
   — What happens to the paper if we place smoked glass or Polaroid in front of the lens?

SECTION C

Energy Education at the Secondary Level

Introduction

Teachers of energy are often times familiar with the fundamental concepts of energy. They are less familiar with conversion technologies and effective energy teaching practice. It would be extremely difficult to produce a comprehensive section dealing with energy conversion technologies. This is not only because of the large volume of available material and the depth of the subject; but also, because the technological advances in some conversion technologies are so rapid that the section would be out of date prior to its publication. It is for this latter reason, coupled with changing economic conditions and consumption patterns, that many commercial publishers have stopped publishing books on energy. The secondary group consequently focused upon producing examples of various effective teaching strategies and methodologies as material which would be of most interest to those who would likely be interested in this section of the volume. The introductory article in this section by the team leader, Tim Hickson, succinctly expresses the fundamental aims of this section's content.

7

Teaching about Energy at the Secondary Level

T. D. R. HICKSON

The King's School, Worcester, U.K.

Energy has always been taught in school science courses. It has, however, normally involved examining events occurring only in the laboratory or classroom. The interaction between science, technology and society is obviously important and will become increasingly so. As a result, in recent years it has become obvious to teachers all over the world that schools have a duty to ensure that their pupils are aware of this interaction and of its consequences. Many of our pupils will go on eventually to become decision-makers or to have an influence on decision-makers or to have an influence on decision-making, if only as voters. Therefore, not only must our pupils be aware of the part played by science and technology in their lives and that of their countries, but they must be aware that decisions on technical matters are not made on the basis of technical information alone but that economic, political, social and environmental factors must also be taken into account.

If these wider issues concerning energy are to be introduced to our teaching then there are a number of problems which must be tackled. Some schools may be fortunate enough to be able to introduce a course on energy or to find a place for the topic to be covered in a Liberal, General or Social Studies course. Most, however, already have a crowded curriculum so there is no opportunity to introduce large chunks of new material. In such schools, energy issues would have to be used to illustrate or to enhance the topics already being taught or to provide a setting for the principles which are to be covered. For example, in physics teaching, when dealing with power calculated from current × potential difference, questions could be set on the energy consumption of an electronic calculator. This could lead on to the cost of using batteries to provide energy and, perhaps, the cost of energy from other sources. A solar powered calculator could be used to open up the whole issue of solar energy and this, in turn, would provide opportunities to do more physics calculations and to explore economic, environmental and social issues. Alternatively, in chemistry teaching, both oxidation and reduction could be dealt with as examples of processes occurring in a wood gasification plant. This too would raise many other issues.

In countries where this type of approach to teaching about energy has been tried, it has been found by the pupils to be highly stimulating. Furthermore, their increased enthusiasm is reflected in a greater motivation to learn generally. Other, less intriguing parts of the subject benefit!

Even when this style of approach to teaching energy topics is accepted as desirable, there remains one major problem. Most teachers left school, perhaps studied their subject at the tertiary level, trained to teach and then went back into the schools. Apart from in their homes, they have little experience of the "real" world outside the classroom, laboratory or lecture theatre. If they are to deal with the wider issues of energy, let alone the other topics covered at this conference, they will need help.

Finding *information* about energy issues is not difficult. It appears in newspapers, in magazines, on the radio and TV: there is a vast amount of literature available, a lot of it free. However, the *techniques* by which the material can be taught, or infused into existing teaching, is less readily available and so the Energy Group has concentrated on the task of assembling such ideas. We have tried to provide a wide variety of examples of the various teaching techniques to illustrate their use in different subjects and in a range of national settings. We hope that those teachers who read what we have produced will feel encouraged to use the ideas, perhaps adapting them to their national setting, and also be encouraged to generate many more examples of their own.

Teaching Techniques

Listed below are some of the techniques which could be used, together with examples.

Techniques		*Examples*
Data Analysis:	Data from published graphs or tables or surveys carried out by the pupils themselves.	Finding local/national/ world fuel consumption trends. Household consumption from energy bills.
Questions:	"What if...?" questions used to raise issues for discussion, for example.	"If the whole world was one big drop of oil how long would it last at our present rate of consumption?"
Pictures:	Thought-provoking or surprising illustrations.	Environmental problem of wind-powered generators indicated by scale diagrams.
Experiments:	Teachers' demonstrations or pupils' experiments at school or at home.	Comparison of using electricity and gas to boil water for a cup of tea.

Techniques		*Examples*
Posters:	As an attractive method of conveying information.	A pupils' survey of energy use by a community.
Games:	Simulated case-studies as decision-making exercises.	Choosing different types of power station. Animal wastes used for fertiliser compared with burning it as fuel.
Visits:	To see science and technology in action and its consequences: to gather real data.	Visiting a coal mine or village bio-gas plant.
Reading:	Gathering information and ideas from a range of different sources.	Comparing articles by opponents and proponents of nuclear power.
Films/Videos:	A visually stimulating way of conveying information and promoting discussion.	"Energy in Perspective" from BP Educational Service.
Audiovisual:	Tape-slide packages are relatively inexpensive ways of presenting information and stimulating discussion.	"Vanishing Forests (India)" from the International Centre for Conservation Education.
Computers:	Will make an increasing contribution to techniques for teaching about energy.	Programs are available which allow energy conservation options for a home to be compared.
Newspaper Cuttings:	Provide a readily available resource covering topical local/national and international issues.	Countless examples such as opposition to hydro-electric schemes, complaints about acid-rain, praise for nuclear power.
Debates:	To teach the need for evidence to support opinions and the skill of organising arguments.	Physics, chemistry, biology, geography, economics students debate a proposal for a tidal power scheme and its likely consequences.
Audio cassettes:	As an inexpensive means of letting pupils listen to "experts" giving their views on energy topics in order to stimulate further debate.	Well-known people discuss "The Energy Crisis" and then pupils are asked questions about what they have heard.
Competitions:	Rouse enthusiasm to learn what is needed to compete and to win!	Using the energy stored in a rubber band to propel a vehicle for the longest distance.

It may be worth pointing out that the skills learnt in this way are then available for other subjects in the curriculum and for later learning.

The Wider Aspects of Energy Issues

It is not always immediately obvious how rich a topic may be in providing opportunities for its inclusion in a range of school subjects. To illustrate this point, one topic was presented, first in its barest form. That chosen was fossil fuel. Many others could have been used.

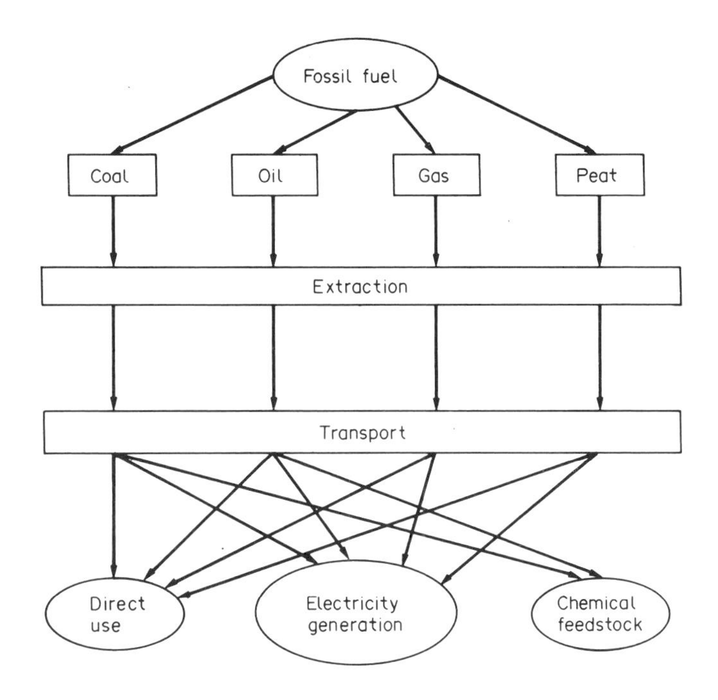

FIG. 1 Fossil fuel (i)

The traditional school subjects were then added to illustrate which parts were usually found a place in the secondary curriculum. The whole topic is rarely to be found being taught in one subject.

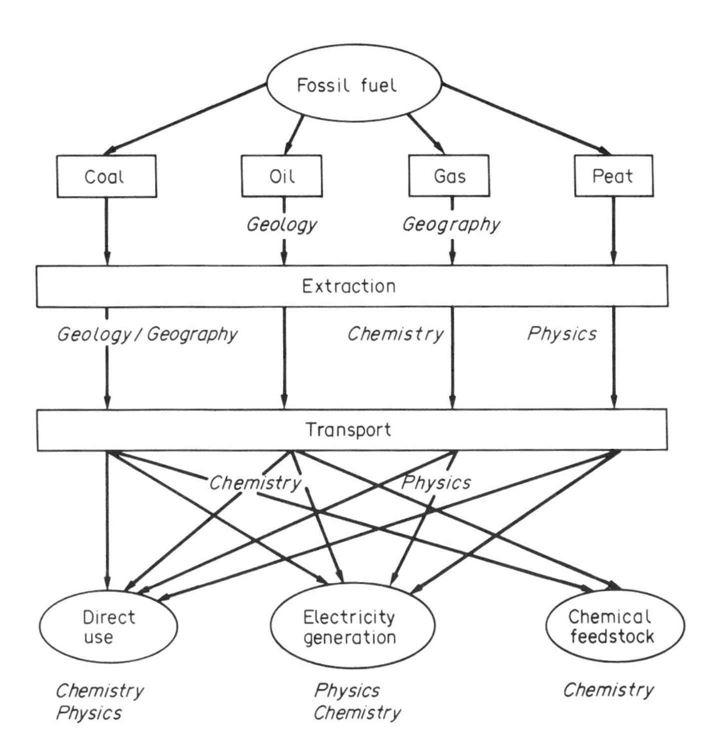

FIG. 2. Fossil fuel (ii)

Then subjects were added which are newer to the curriculum. Indeed they may not all be taught in any particular school or college.

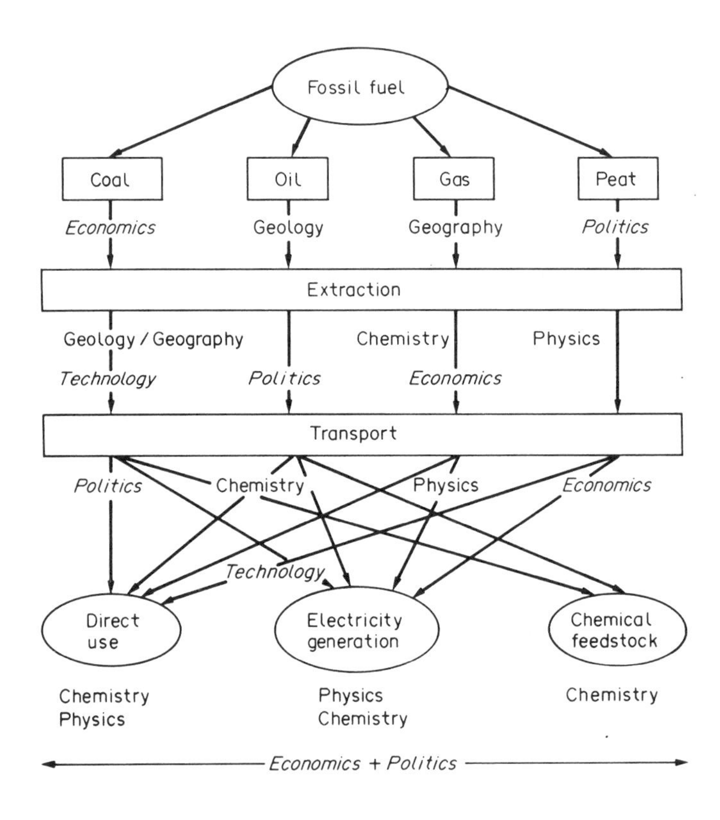

FIG. 3. Fossil fuel (iii)

Finally, many other aspects of the topic were added. These usually have no exclusive place in any one curriculum subject. Increasingly they appear, with a different flavour, in several subjects.

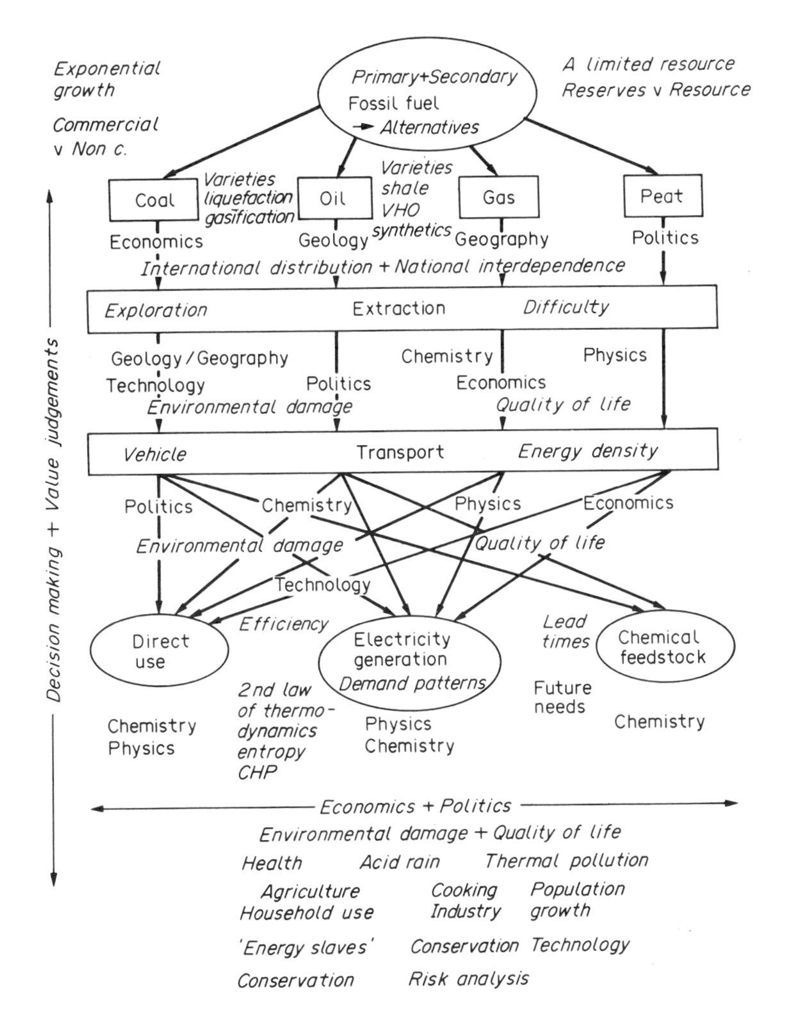

FIG. 4. Fossil fuel (iv)

If a particular energy topic crops up in the teaching of a number of subjects, not only will this help to provide a balanced view but it will emphasise the fact that *decisions* on energy issues require contributions from a wide range of disciplines.

8

Strategies for Promoting an Interest in Energy Issues at the Secondary Level

T. D. R. HICKSON, J. L. LEWIS and A. B. PRAT

USE OF A SURVEY

A. B. Prat
Torino, Italy

Three questions are put:

1. Which of these named kinds of energy are the **two** most widely used today?
   (In Europe, people could select from oil, gas, coal, nuclear, solar, hydro and wind.)
2. Which will be the two most important energy sources in the year 2000?
3. If **you** could decide, what would be the two main kinds of energy used in the year 2000?

Pupils can either ask these questions of themselves or they can conduct a survey amongst others. In either case they can then make histograms.

Their histograms in response to the first question can then be compared with the actual situation. This is usually revealing as often pupils have little appreciation of what are the main sources of their energy supplies.

The histograms in response to the second question can then be compared with the scenario for the year 2000 produced by expert opinion. The response to the third question can lead to lively discussion.

Such a survey was carried out in Europe by Shell International and the results were reported at an ICASE conference in Monte Carlo in 1981 by C. D. van Lennep. His results are shown below.

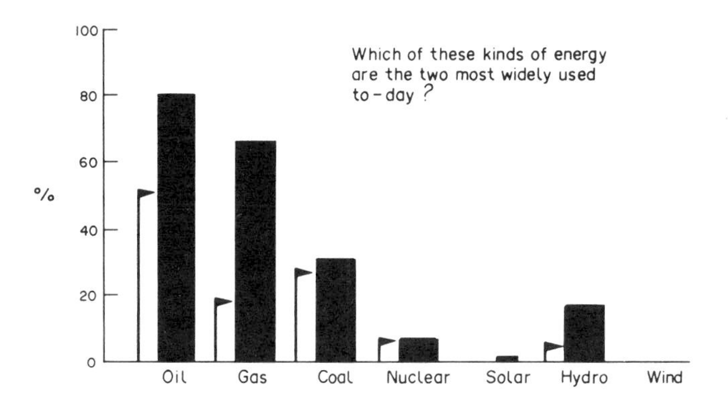

The bars show the spread of answers to the question.
The flags show the actual contributions of the various energy sources in Europe in 1980.

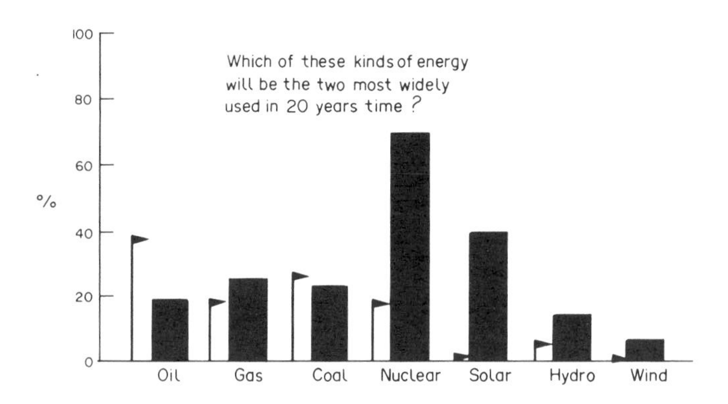

The bars again show the spread of answers to the questions.

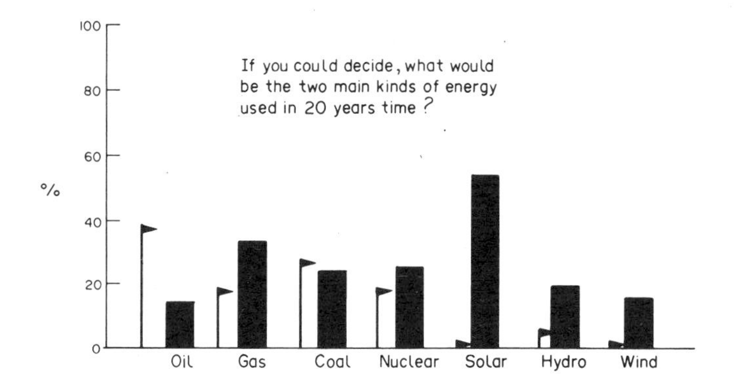

The flags indicate what a group of experts saw as a probable future energy balance which is technically achieveable.
The results are certainly significant showing how the aspirations of the public differ from what is in fact possible.
FIG. 1 Graphs reproduced with permission from the International Council of Associations for Science Education

COLLECTING AND ANALYSING DATA ON FAMILY USE OF ENERGY

J. L. Lewis
Malvern College, U.K.

A useful exercise to stimulate interest and to help pupils to become more aware of the importance of different fuels and the uses to which they are put is discussed in detail in the Energy section of the Science in Society Project[1] published in the U.K. It is also useful in getting pupils aware of the different units in which energy is measured.

Pupils find out how much energy their family has purchased in a year: this usually requires the sympathetic co-operation of parents! If possible, the information should include details of food consumption.

Typical figures for a U.K. family of four are:

"Fuel"	Amount	Unit	Number of joules
Electricity	4000	kWh	1.4×10^{10}
Natural gas	1500	therm	15.9×10^{10}
Coal	1	tonne	2.8×10^{10}
Wood	0.1	tonne	0.1×10^{10}
Petrol/Gasoline	2000	litre	8.0×10^{10}
Food	4,000,000	kilocalorie	1.7×10^{10}
		Total	3×10^{11}

Conversion of the figures to the single unit of the joule prompts questions which lead to useful discussion. Also, it can be seen that, very roughly, the average

individual use of energy is 10^{11} joules per year. This in turn can lead to a comparison between countries and the graph of energy *per capita* against gross national product shows the enormous difference between certain developing countries and those with a more sophisticated economy.

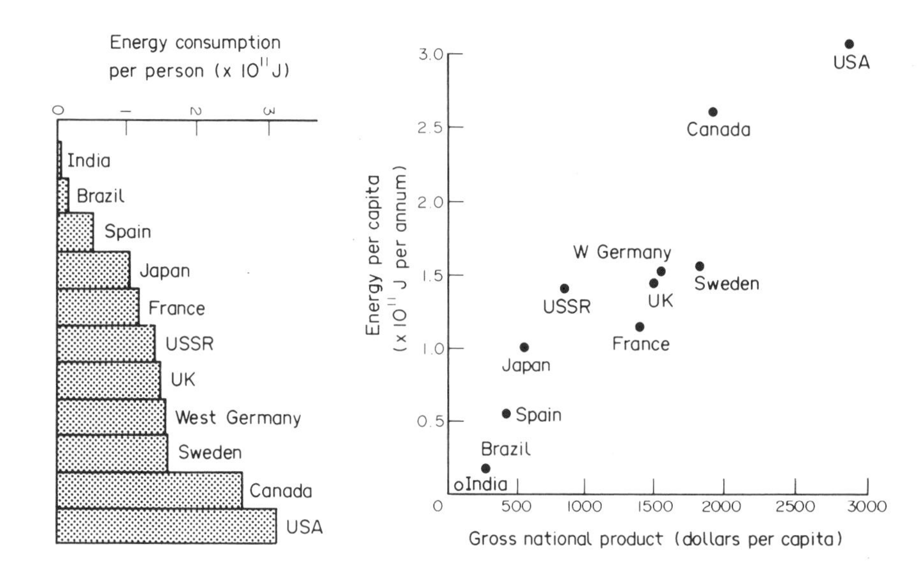

FIG. 2

Reference

1. The science in Society material is obtainable from the Association for Science Education, College Lane, Hatfield, Herts AL10 9AA, U.K.

COMPARISON OF DIFFERENT NATIONAL DEMANDS FOR COMMERCIAL FUEL

T. D. R. Hickson
King's School, Worcester, U.K.

To promote an interest in and discussion about the different patterns of demand for commercial fuel, as well as such issues as the distinction between commercial and non-commercial fuels, the consequences of different climates and the requirements of nations with different degrees of technological advancement, a poster display or an overhead transparency with the following information is useful.

The information appears in "North–South: A Programme for Survival" (often called the Brandt Report), published by Pan Books, 1980.

1	North American
2	Germans or Australians
3	Swiss or Japanese
6	Yugoslavs
9	Mexicans or Cubans
16	Chinese
19	Malaysians
53	Indians or Indonesians
109	Sri Lankans
438	Malians
1072	Nepalese

The commercial energy used by one North American in a year is (on average) equal to that used by two Germans or Australians and by an increasing number of people from other nations.

Pupils find the information surprising and the list makes even more impact if it is gradually revealed starting at the top.

TOTAL ENERGY PER PERSON

J. L. Lewis
Malvern College, U.K.

Another way to promote awareness of the disparity between nations is to consider the total national consumption of primary fuels.

As a preliminary, a student might be timed running up a flight of stairs. A student of mass 50 kg might take 5 sec to run up a flight 10 m high. The energy transfer would be 5000 J and the power would be 1000 W or 1 kW. But for a working day of 8 hr, it would not be possible to maintain that rate—a tenth of it might be more appropriate, in other words a power of, say, 100 W. At that rate, the work done in 8 hr would be $100 \times 8 \times 60$ J or about 3 MJ.

The next figure to take is the total national consumption of primary fuels. These figures are usually available from government statistics or can be put together from different sources. For the U.K., an approximate figure for 1980 was 10^{19} J. With a population of 55×10^6, the annual consumption per person can be worked out as about 1.8×10^{11} J. From this, by dividing by the number of days in the year, we get that the average energy person per day is about 500 MJ.

If we were to take a day's work done by a "slave" as 3 MJ, as worked out above, it means that every man, woman and child in the U.K. has the equivalent of 167 slaves working for them.

Of course this is only a crude calculation, but it makes a valuable educational point. The number of "slaves" available per person can be worked out for both developed and developing countries and the contrast is considerable.

A further point can be made by considering what is paid for some forms of energy. For example, 1 kWh is 3.6 MJ and costs only a few pence (or a few cents), showing that the "slaves" are paid at a very low rate. It is therefore obvious that those countries which have access to large sources of cheap energy are bound to grow at a very different rate from those developing countries which do not have such energy available.

There is value in these order of magnitude calculations, despite their crudity, as well as promoting awareness of energy problems in the world.

9

An Outline Programme for Teaching about Wind Power

ROSE MALONE

Trinity College, Dublin, Republic of Ireland

M. P. GOVINDARAJAN

Chalavara High School, Kerala, India

Introductory Work on Wind

Reference to solar origin, atmospheric pressure (simple demonstrations), use of instruments (barometer and anemometer).

Collection by students of data on meteorology (analysis, interpretation and application), types of wind (relate to geography).

Use of wind as a direct source: sailboats (calculations, design comparisons from different cultures); simple wind machines (fans, kites, gliders, etc., calculations); history of the use of wind energy (available technology and social conditions).

Conversion of Wind Energy to Other Forms

1. Direct to mechanical energy
   — history
   — calculations
   — models
   — practical activities
2. Conversion to electricity
   — calculations
   — models and design problems
   — practical activites
   — use of materials (blades and towers)

Social, Environmental and Feasibility Aspects of Wind Power

Advantages and disadvantages — siting problems

— need to store energy and methods of storage
— noise
— appearance

Decision-making exercises

The above outline is one way in which this topic might be covered, quite exhaustively. Most teachers would wish to spend only a short time dealing with this form of energy conversion. For them, perhaps a brief introduction followed by one of the activities described below might be appropriate.

1. The Power of the Wind

Newspaper Cuttings

As an introduction to the idea of the wind as a source of power, students should collect pictures and articles illustrating the effects of storms, cyclones, hurricanes or whatever is appropriate to their locality.

Discussion may draw out the facts that a great deal of energy is available, that there is not often a steady, reliable source of supply, that the strongest winds may come at a time when the need for energy is greatest but that we may not be able to harness the strongest winds.

2. Energy from Wind: Simple Meteorology

Data Analysis

Students can be asked to collect weather data, such as wind speed temperature, atmospheric pressure. They can be asked to analyse, interpret and apply information, for example, to graph wind speed against time of day or time of year. A wind speed of 5–6 km/hr is required for the generation of electricity. Questions such as "At what time of day is there enough wind to make electricity where you live?", "On how many days last month was there enough wind to make electricity?" can be posed and answered by direct observation. A simple anemometer can be constructed by cutting two hollow plastic spheres in two and cutting a square notch on each diameter. These are mounted with glue on square aluminium rods of corresponding dimension. The rods are joined at their centre by drilling and passing a screw through the hole. The entire apparatus must be stable—inequalities can be corrected by adding small amounts of glue. The apparatus is mounted on a small electric motor, connected to a milliammeter.

The device can be calibrated by taking it out in a motor car on a still day—when the car is travelling at 30 km/hr, the anemometer is turning at 30 km/hr. Speeds of approximately 40 km/hr should not be exceeded.

References

1. *Alternative Energy Sources: "Experiments you can do"*. Thomas Alva Edison Foundation, Cambridge Office, Plaza, Suite 143, 1820 West Ten Mile Road, South Michigan 48075, U.S.A.
2. *Il Problema dell'Energia*. Zanicelli Editore, Italy (Construction of anemometer).

3. Wind Energy—Conversion to Electrical Energy

Experimental Investigation

To investigate the effect of the angle of the blades of or propeller on the generation of electricity.

A simple propeller can be constructed from balsa wood as follows:

A 3-cm square is cut from balsa wood, approximately 1 cm thick, to act as the spinner. A small hole is cut in the centre to take the motor shaft. A groove 2 mm deep is cut on each side of the spinner to take the propeller blades. The blades are cut from balsa wood approximately 2 mm thick. Each blade is 3 cm wide by 6.5 cm long. The blades are glued into the grooves, making sure that they are angled in opposite directions.

The propeller is pushed on to the motor shaft. The leads of the motor are connected in parallel with a 100-ohm resistor and a voltmeter. The propeller assembly is held in front of a fan and the voltage reading recorded. A second propeller is constructed, with the grooves in the spinner cut at a smaller angle and the experiment repeated.

Many questions can be posed to students.

1. Which angle produces most electricity?
2. How can efficiency be improved?
3. What is the effect of altering the size of the blades?
4. Does the material of which the blades are constructed have any effect?
5. What will be the effects of scaling up the model?
6. Compare four-bladed and two-bladed propellers.
7. What is the effect of cutting the wind speed in half?

Reference

Paul Oei and Eugene Sorensen, *Projects and Experiments in Energy*. National Energy Foundation, U.S.A.

4. Wind Generator Design

A Design and Construction Competition

Students may be divided into groups and challenged to design and then construct the most efficient model wind generator from simple materials.

The use of groups gives rise to constructive discussion.

Constraints may be worked out by the class as a whole. For example, they might decide:

(i) what materials may be used,
(ii) how much material may be used,
(iii) how will efficiency be measured.

5. Aesthetic Considerations of Use of Wind Generators

Posters and Visits

Students should be organised to make a large collage poster of pictures of traditional windmills. This will remind them of the images with which they are familiar. Then get them to add, **to scale,** pictures of modern wind generators. The contrast is likely to be found impressive.

As an alternative, an OHP transparency could be prepared so that first a traditional windmill appears on the screen, then increasing large wind generators are revealed. An added point is made if the final picture that appears is of a powerline pylon. This, which is often considered to be an eyesore, is considerably smaller than designs for some modern wind generators.

If possible, this activity should be combined with a visit to a wind generator.

6. The Alternative Energy Project

A Decision-making Game

This project is published by the Association for Science Education in the U.K. but it could be adapted to suit most localities in the world. It is a simulated case study concerning, in this case, an island off the coast of Scotland. Should it use peat, solar power, wind power, tidal power or hydropower or some combination of these to satisfy its energy needs for the next fifty years?

Students are divided into groups so that each group is responsible for investigating the economic, technical, environmental and social aspects of one possible source of electricity. Unusually for a decision-making exercise, instead of arguing for their particular resource in competition with the others, they have to co-operate so that the group as a whole can produce what appears to be the best rolling programme for the future. That there is a variety of possible solutions helps to make this a particularly thought-provoking activity.

References

1. *Energy—a Guidebook,* Janet Ramage, Oxford University Press, U.K., 1983.
2. *Energy—Crisis or Opportunity,* Diana Schumacker, Macmillan, U.K., 1985.
3. *Alternative Energy Sources—for the centralised generation of electricity,* R. H. Taylor, Adam Hilger, U.K., 1983.
4. The Association for Science Education, College Lane, Hatfield, Hertfordshire, U.K.

10

Teaching about Solar Energy

HANNA GOLDRING

Weizmann Institute of Science, Israel

J. KUMI

Academy of Arts and Sciences, Accra, Ghana

In introducing the subject it would be good to stress the fact that almost all energy resources are derived from solar radiation. One could show, for example, a diagram similar to Fig. 1 to illustrate this. The present section deals with the **direct** utilisation of solar energy.

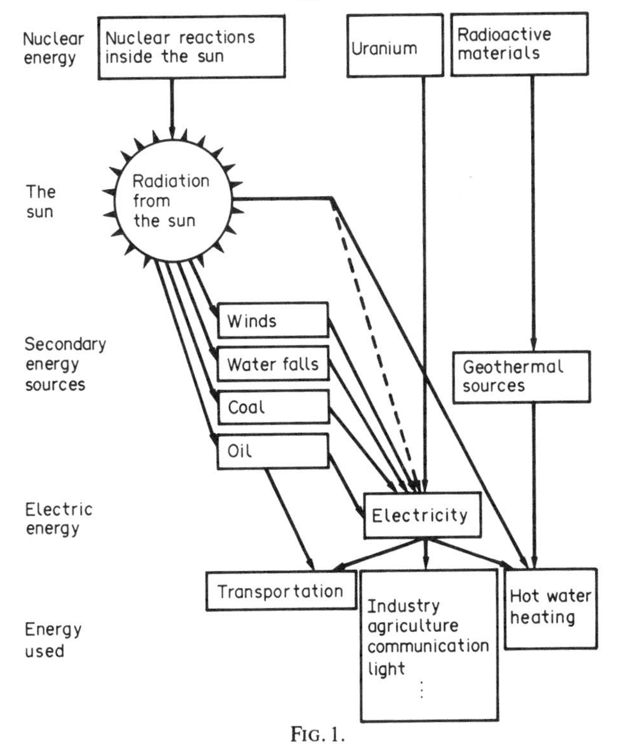

FIG. 1.

Experiments and Activities

One could start with one or two experiments demonstrating the conversion of solar radiation into other forms of energy.

Solar Music

A music box operated by sunlight falling on photovoltaic cells, playing: "You are my sunshine".

Crookes Radiometer

Another good introductory experiment.

Inverse Square Law

Use photovoltaic cells, a light source and a milliammeter to show that the energy radiated by a light source hitting unit area falls off as $1/r^2$.

Energy from Solar Cells

Compare a calculator (or a watch) powered by solar cells with an equivalent device powered by batteries. Thus the energy available from the solar cells may be deduced.

Determination of K

It is important to show that K, the radiative power per unit area received by the Earth is very small; this fact helps to explain why the utilisation of solar energy is so expensive. (From K one can also work out the total power emitted by the Sun.)

The following method is due to U. Ganiel and O. Kedem (see Solar energy—how much do we receive? *The Physics Teacher,* Dec. 1983, pp. 573–5).

The experimental setup

Use two identical rectangular aluminium blocks (2 × 2.5 × 4 cm). A hole is drilled through each block, and a resistor (10 Ω, 2W) is embedded in each hole. Each block is mounted on a styrofoam base, and both blocks are mounted on a solid strip, from which they are isolated by their bases. A second hole in each block is filled with plasticine and a thermometer is pushed into it, so that it is in good thermal contact with the aluminium block. The upper face of each block is blackened using soot from a candle. Directly above each block there is a screen made of a rectangular piece of styrofoam, with a removable part directly above the block. A thin rod is mounted on the solid strip holding the blocks, perpendicular to it, and is used to ensure that radiation from the Sun will hit the black surface at normal incidence. The arrangement is shown in Fig. 2.

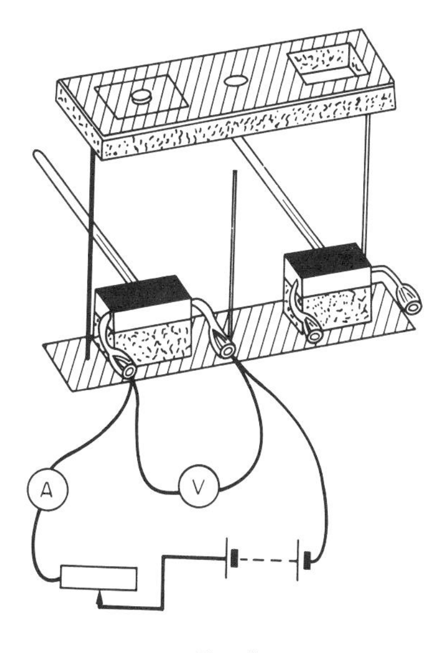

FIG. 2.

The measurement proceeds as follows:

(a) One of the covers is removed, exposing the surface under it to radiation from the Sun. Students watch the temperature rise as measured by the thermometer. Typically, the temperature rise observed is 6.8 °C. Once the temperature is constant, dynamic equilibrium has been reached, the energy absorbed during any time interval at the surface equals the energy lost by the block by all loss mechanisms.*

* The main loss mechanisms are radiation and convection by air at the surface of the block. However, none of these details matters for the purpose of the experiment.

(b) The second block is shaded from the Sun by the cover above it. The resistor embedded in the block is connected through a rheostat to a voltage source (three 1.5 V batteries are appropriate) and the electric current causes heating of the block.

By changing the current with a rheostat, and noting the temperature of the block, one can easily reach a situation where the temperature of the block is equal to that of the other block, and remains constant. Hence, steady state is established here too. One now records the voltage across the resistor (V) and the current through it (I).

The loss mechanisms in both blocks are identical, so the rates of energy loss are the same for both blocks. Therefore, the power inputs into both blocks are also equal.

(c) The equality of both input powers yield a direct measurement of the solar radiation power hitting the surface. We have: $P = VI$ for the power dissipated in the shaded block, and this equals the power absorbed from the Sun by the other block. The irradiance (power per unit area) is therefore

$$E = \frac{VI}{A}$$

where A is the area of the exposed surface.

Inverse Temperature Gradient in a Solar Pond

Demonstrate creation of inverse temperature gradient in solar pond (high temperature at the bottom, low at the top).

(i) Take two plexiglass tanks (40 × 20 × 60 cm), insulate them by means of styrofoam.

(ii) Drill holes in the tanks at the bottom, in the middle and near the top and insert thermometers.

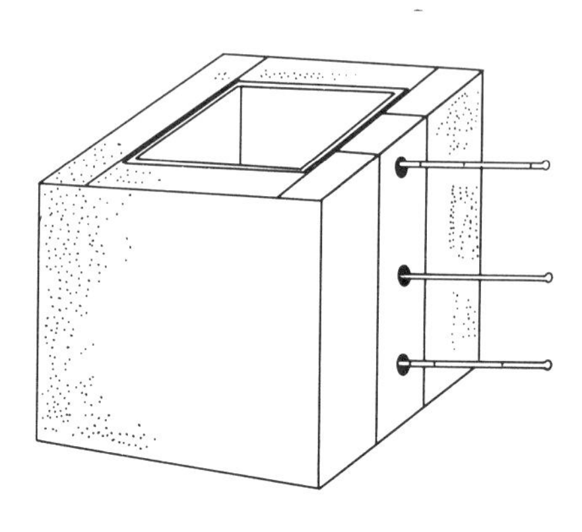

FIG. 3.

(iii) Fill one tank (A) with a uniform solution of salt in water, and the other (B) with a salt solution of varying concentrations, high at the bottom and low at the top.

(iv) Take care that both tanks contain the same total amount of salt and contain the same volume of solution. When both tanks are left to stand in the Sun for a few days, equilibrium is established and one observes the inverted temperature gradient in tank B, and no temperature gradient in tank A.

Model of a Solar Panel

Construct a model with a tank to demonstrate the use of a panel to heat water.

Experiment Kit

A kit which enables one to perform 25 experiments has been developed by, and is available from V. B. Kamble and S. Premchandran, The Vikram A Sarabhai Community Centre, Navrangpura, Ahmedabad 38009, India. An instruction booklet is also available.

Some of the experiments that can be set up with this kit are:

Heating of water in containers exposed to sunlight.
Distillation using solar energy.
Radiation heat loss.
Sun tracking.
Greenhouse effect.
Parabolic concentrator for water heating.

DEMONSTRATION TO SHOW WHY SOLAR COLLECTORS ARE NOT ALWAYS HORIZONTAL

Don Kirwan
University of Rhode Island, U.S.A.

Materials: Photovoltaic cell connected to a milliammeter; light beam.

Procedure: Initially orient the photovoltaic cell such that the surface is perpendicular to the light beam. Observe the reading on the meter. Slowly vary the angle between the cell and the light source. The meter reading will decrease from a maximum to zero according to $I = I \cos$ (angle).

Lesson: To receive the maximum account of solar radiation, the collector should be oriented such that the surface (or aperture) is perpendicular to the Sun's rays.

Exercise: Have students devise a simpler method of showing this same phenomenom. *Hint*: This can be done with a piece of material with a hole in it. As the material's orientation is varied, the light emerging through the hole is spread out over a smaller area than the size of the hole.

A RESOURCE QUESTION ON ENERGY POTENTIAL

Don Kirwan
University of Rhode Island, U.S.A.

The question: Can solar energy satisfy the energy needs of the U.S.A.?

The U.S.A. has a land area of 9.36×10^6 km^2.

Assumptions:

1. Average power received from the Sun (U.S. average) of 180 W/m^2.
2. 10 per cent direct energy conversion efficient ratio.

In one day energy/m^2 = 180 W/m^2 $\times$ 24 hr = 4.32 KWh/m^2.

Thus the energy falling on the U.S.A. in one day is:

energy/day = 4.32 KWh/m^2 $\times$ 9.36×10^{12}m^2 = 40.4×10^{12}KWh/day

and in one year we have:

energy/year = 40.4×10^{12} KWh/day $\times$ 365 days/year = 1.5×10^{16}KWh/year at 10 per cent conversion, this is 1.5×10^{15}KWh.

Assume that the U.S.A. needs 1.0×10^{13} KWh, then we conclude that we have *PLENTY to SPARE!*

This implies that we only need to use 0.6 per cent of our area for solar panels. This translates into an area of 24,500 square miles. This is an area slightly larger than the state of West Virginia. The consequences of covering this area with solar panels makes a good exercise for the students. If our energy needs are larger, then the consequences are even more devastating.

Activity—Newspaper Cuttings

Have students cut out relevant articles from newspapers, and discuss their contents in class. The articles should be examined critically for any errors or misrepresentations. Discuss them with students.

Decision-making Games

Calculate the cost of hot water per month for a family using flat plate collectors, oil or electricity.

Problems Arising in the Generation of Electrical from Solar Energy

A suitable teaching strategy would be to bring these out by a discussion with the students. They should have all the basic information needed. The common problem of electrical power generation is the high cost, because the density of solar radiative power falling on the Earth is so low. Thus very large areas are required for collecting a sizeable amount of power, and land is usually expensive. Another problem is that of **storage,** since sunshine is intermittent and totally absent at night.

The methods of generation of electrical power used at present are mainly:

1. Photovoltaic cells.
2. Solar central receiver.
3. Solar pond power plant.

They are described in the literature, e.g. *Alternative Energy Sources* by R. H. Taylor, Adam Hilger Ltd, Bristol, 1983.

The solar pond power station has built-in storage capacity since the warm layer at the bottom stays warm for some time at night, depending on how much energy is used.

It is a common misconception that generation of electrical energy from solar radiation is risk free. Solar power stations occupy large areas; the larger the area of an installation the bigger the chance of accidents. Usually the mirrors or collectors are placed high above the ground, this also adds to the chance of accidents. In a solar central receiver a mirror which gets out of control and reflects radiation to the wrong place, can be lethal.

It should be pointed out that emissions from thermal power stations cause heating up of the surface of the Earth, which may cause the melting of ice caps near the Poles, and flooding. A solar power station is free from this effect. The students should discuss this matter.

Teaching Techniques in Rural Areas

The following current usages of solar energy can be demonstrated on a small scale in schools.

1. Heating water—just leave vessel in the Sun.
2. Salt production by evaporation of salt water in the Sun.
3.* (a) Drying of various crops, pepper, cassawa chips, cocoa, maize, etc.
   (b) Preserving fish by soaking it in brine, then drying it in the Sun.
4. Construction of small mud-houses, ovens and stoves: Wet clay is left to dry in the Sun after being moulded into these structures.

* *Trap the Energy of the Sun,* Science and Technology Education in Philippine Society. From Science Education Centre, University of the Philippines, Diliman, Quezon City, Philippines.

THE MEDITERRANEAN—DEAD SEA PROJECT

H. Goldring

Weizmann Institute, Israel

This project is a hydroelectric power plant which will make use of the difference in heights, about 400 m, between the Mediterranean and the Dead Sea. A pipeline and tunnel is to be constructed between the two. In this canal water will flow from the Mediterranean to the Dead Sea. If too much water is let into the Dead Sea, its level will go up, and the difference in heights will go down. The amount of water we can let flow into the Dead Sea (per year) without increasing its level is fixed by the rate of evaporation of water from the Dead Sea, and this depends on solar radiation. One can therefore consider the Med-Dead Project as a solar operated device.

The amount of water planned to flow in the canal, was supplied in the past by the Jordan river flowing into the Dead Sea—about 1600×10^6 m^3 per year. With this rate of flow the Dead Sea level remained constant. As a consequence of irrigation projects established by Israel and Jordan, this flow has stopped almost completely. The Med-Dead project will therefore supply water to the Dead Sea at the rate of flow of the Jordan and its tributaries.

Exercise

Calculate the power output of this hydroelectric solar plant, assuming 90 per cent efficiency of conversion.

Discuss how this plant could be operated to obtain more power during periods of peak demand.

Discuss possible implications of this canal on the national economy. (Increased power production, topping up solar ponds at the Dead Sea, cooling of conventional or nuclear power stations to be constructed in the future.)

Booklets from the Philippines

There are a number of excellent illustrated booklets produced by the Institute for Science and Mathematics Education Development, University of the Philippines, Vidal Tan Hall, Pardo de Tavera St., Diliman, Quezon City, Philippines. These were brought to the conference by Vivien Talisayon. Each covers a teaching module containing experiments—most using easily constructed apparatus—information, activities and questions. Three are relevant to a study of solar energy.

Solar Cooking

This module describes the construction, use and principle of low-cost solar cooker made of carton and plastic. The temperature inside several boxes of different materials under different conditions is compared. The temperature inside and outside the cooker is also compared under different time intervals. Students use the solar cooker to boil eggs, yams, and banana (plantain variety).

Potable Water from Sea Water by Solar Distillation

This module explains the construction and use of a solar still, a device for distilling sea water using solar energy. The solar still is constructed using wood and glass or plastic.

Trapping the Energy of the Sun

The construction and operating principle of a solar dryer are explained. The students are shown how to build a low cost solar dryer and to compare the drying time in the dryer and in direct Sun drying, and the temperature of the air inside and outside the dryer. Students also measure the moisture content of the palay being dried in the dryer and directly under the Sun. The students are also asked to make some changes in the basic design of the dryer and observe the effect on the drying time. Finally, the students get to build a dryer three times bigger than the first one with a capacity ten times greater.

Further Suggested Activities

The Sun as Heat Source

(a) Focusing solar radiation by a lens or mirror on to water in blackened calorimeters and polished calorimeters. Heating curves can be plotted.
(b) Painting or polishing aluminium squares, mounting them in an insulation box (e.g. a seed tray) and exposing them to the Sun's rays. In a very short time, temperatures can be measured and the surfaces compared (Fresnel lenses originally intended for overhead projectors are useful for concentrating sunlight).
(c) Melting of ice cubes can be used to give quantitative weight to comparisons of solar concentrations.

Sunlight and the Processes of Science

Projects may be assigned to students to enhance their appreciation of the value of a quantitative approach in a well-designed experiment on solar energy. Projects provide one of the best means by which children and students learn how to behave like scientists and can learn the joys and sorrows of deciding for themselves and putting up with the outcomes.

Social Factors

Discussion

Students may discuss and put down points on:

(a) the effect of the use of the solar energy on the society,
(b) the constraints which are put upon the use of solar energy by social factors.

Debate

Two groups of students can debate on the following:

(i) Energy supplies decide the comfort of individuals and the growth of nations more than any other factor.
(ii) Solar energy is a far better source of energy for human use than nuclear energy.

Books

1. *Practical Physics, The Production and Conservation of Energy,* J. F. Milligan, McGraw Hill Book Co., 1980.
2. *Energy Insights from Physics,* P. Di Lavore, Wiley, 1984.
3. *Energy—Its Physical Impact on the Environment,* D. W. Devins, Wiley, 1982.
4. *Energy for the Future,* John L. Roeder, Calhoon School, N.Y.
5. *Energy Environment Source Book,* NSTA, 1975.
6. *The Health Hazards of Not Going Nuclear,* P. Beckmann, Golem Press, Boulder, CO, 1976.
7. *Solar Energy,* J. I. B. Wilson, Wykeham, Taylor and Francis, U.K., 1979.
8. *Energy,* Readings from *Scientific American,* W. H. Freeman and Co., 1979.
9. *Before It Is Too Late,* B. Cohen, Plenum, 1983.
10. *Science in Society,* Project Director John L. Lewis, Heinemann Educational Books, ASE, U.K., 1981.
11. *Sun Traps,* John Elkington, Penguin Books, U.K., 1984.
12. *Alternative Energy Sources for the Centralised Generation of Electricity,* R. H. Taylor, Adam Hilger Ltd., Bristol, U.K., 1983.

A survey of curricula dealing with solar energy training in various countries is to be found in a report sponsored by UNESCO in co-operation with the National Institute for Higher Education, Limerick, Ireland. The survey was edited by J. C. McVeigh.

11

Teaching about Hydro-electric Power

E. LISK
Sierra Leone Grammar School, Freetown, Sierra Leone

T. SUBAHAN
University Kebangsaan, Malaysia

What follows is a set of activities which could make a teaching programme. Naturally teachers will wish to devise a scheme to suit their needs. This example has been aimed at pupils aged 12–14, but many ideas can be used at other levels.

Objectives

As a course on energy for the lower secondary level (12–14 years) the following are considered useful:

1. Pupils should be able to identify running water as a source of energy.
2. To describe the energy changes in a water-wheel and a hydro-electric plant.
3. To construct a water-wheel and a hydro-electric plant model.
4. To calculate energy changes in a hydro-electric plant model.
5. To discuss the natural assets which favour construction of a hydro-electric plant.
6. To discuss the threat a hydro-electric plant poses to ecology, health and agriculture.

Activities

1. *Water cycle.* A poster should be prepared to show the Sun as the source of water energy.
2. *Rivers as sources of energy.* Pupils should be asked to consider how rainfall raises the water level of rivers and how rapids are formed when a

river bed drops sharply from a highland to a lowland. Suggestions for harnessing the energy of rivers would then be discussed.

3. *A water-wheel.* A picture of a water-wheel could be shown so that pupils understand its working principles.
4. *Demonstration.* To show how the translational kinetic energy of water can be used to rotate a wheel, water, which has been allowed to fall through a tube, turns a home-made water-wheel.
5. *Pupils' exercise.* Pupils able to cope with the physics could be asked to calculate the energy changes in a simple system such as that illustrated below.

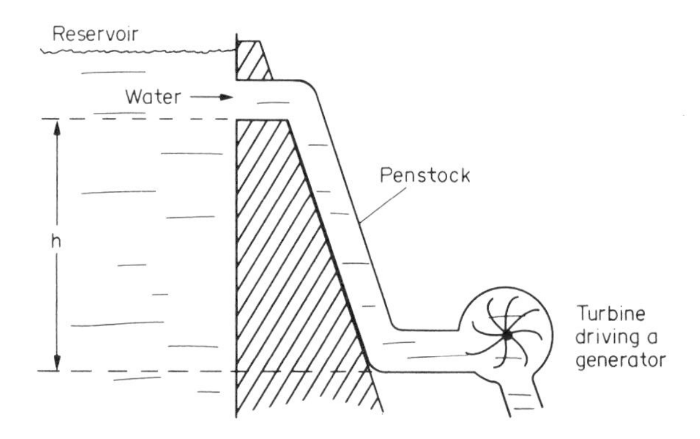

Q1. What are the energy changes that take place?

Q2. If a mass m of water falls through the penstock (tube), what is its loss of gravitational potential energy? *(mgh)*

Q3. If that mass of water takes a time t to flow through the tube, what power is delivered to the turbine? *(mgh/t)*

Q4. What factors would limit the efficiency with which this power is converted to electrical power?
The actual efficiency of this conversion may be 90 per cent.

Q5. If such a power station had to deliver 200,000 W, what would have to be the power delivered to the turbine? (about 220,000 W)

Q6. Suppose the height h was 300 m, what would be the rate of flow of water in kilograms per second? ($^m/_t$ = 75 kg/sec)

In the light of this calculation, pupils might consider the potential of any local sites for the production of hydro-electricity.

If the pupil's locality cannot provide high reservoirs but has fast-flowing rivers instead, then the calculation could be modified appropriately.

6. *The motor car alternator.* In order to stress the feasibility of the hydro-electric project the working principle of the alternator in a motor car

could be discussed. Pupils would be able to see how the fan belt turns the armature of the alternator. The leads should be taken off the alternator and connected to a voltmeter. Pupils will be able to see that as the power of the engine is increased by increasing the pressure on the accelerator pedal, the armature of the alternator turns faster and the reading of the voltmeter increases.

7. *Newspaper cuttings.* Pupils could collect or be shown newspaper cuttings which expose the hazards of harnessing rivers for hydro-electric energy.

8. *Decision-making game.* Pupils could be given an exercise to decide where to site an electric power station, and why a hydro-electric power station is preferable to nuclear power station, oil-fired station or a coal-fired station. The format of the "Power Station Project" from the Science and Society (U.K.) Course can be modified for this purpose. (The U.K. Institute of Electrical Engineers has produced a quite elaborate game dealing with hydro-electric power called Hydropower.) Contrary to the situation in Britain, there exists a number of fast flowing rivers in Malaysia, Indonesia and many parts of Africa. The aim of the game is to get the pupils to understand how to make wise decisions based on scientific considerations. Pupils will be asked to consider the hazards which a hydro-electric project pose to agriculture, the environment, the ecology and to social and moral values. They must be able to find ways of reducing these hazards so that the advantages of constructing the hydro-electric plant outweigh these disadvantages.

9. *Visits to hydro-electric plants.* Visits to hydro-electric plants would be a valuable experience for the pupils.

10. *Lectures.* A doctor, an environmentalist, a trade unionist, an agriculturalist and a politician could be invited to give lectures on the effect of building a hydro-electric plant in their area.

11. *Assignments.* Pupils could be asked to list (a) all the diseases such as bilharziasis, river blindness, sleeping sickness and malaria which are related to a hydro-electric plant; (b) the programmes which the government has embarked upon to eradicate them. Pupils might also be asked to draw bar charts of total agricultural produce and total rainfall for that area over a period of ten years—five years before the hydro-electric station was built and five years after its construction. They should also be asked to discover any other consequences that followed the construction of such a power station.

12

Teaching about Geothermal Energy

M. KOREK

Rowdoh High School, Beirut, Lebanon

A Possible Programme

This is intended as an example only, for teachers will naturally wish to devise a programme to suit their own needs.

Experiments

A quantity of water is poured onto a heated piece of metal (formation of steam).

Use of the steam.

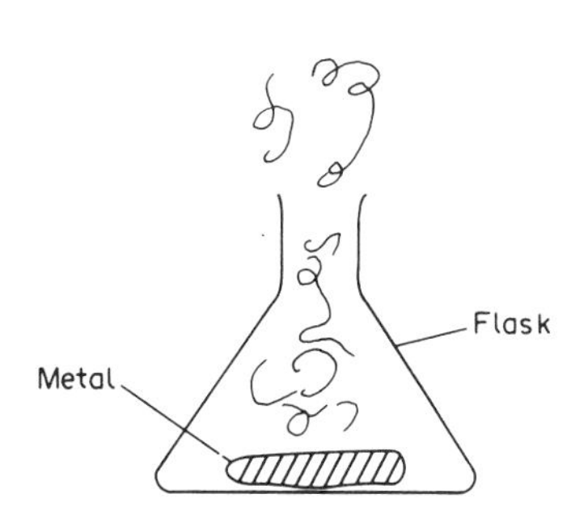

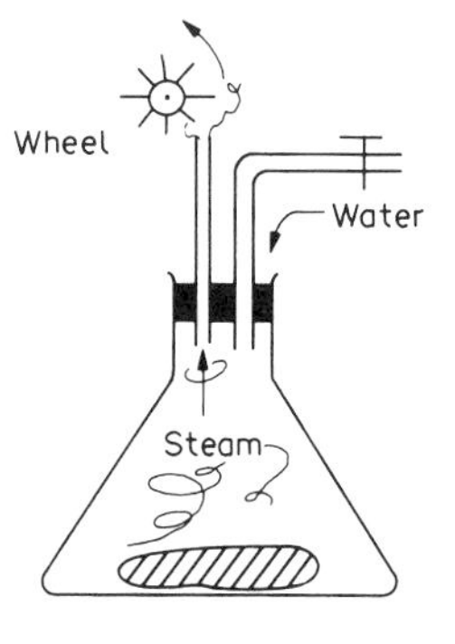

The wheel might be linked to a generator to light a lamp.

Information about the Earth

— We can ask the students to collect some information about the inner Earth from library books. The geographers might be brought in to help here.
— Film loop (few minutes) about volcanoes.

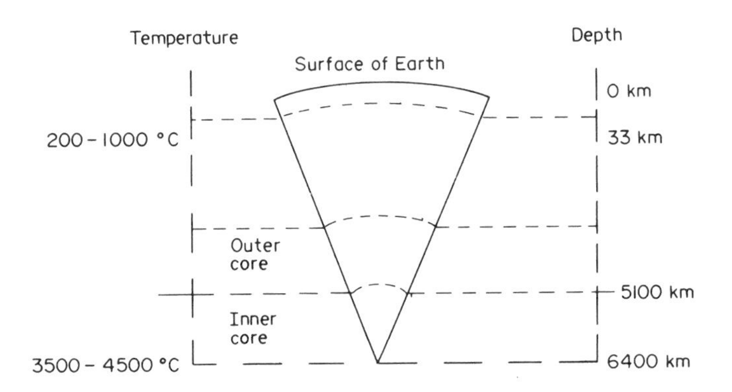

— On a transparency (over head projector): positions of volcanoes and where we can use the geothermal energy on the surface of the Earth.
— An OHP transparency would be useful here.

Drilling

Two holes are drilled in the Earth to reach the hot rocks.
— Film loop (few minutes) showing a drill in operation. (*Remark:* the depth of the holes is restricted because of technical problems as the weight of the cable. For great depths the diameter should be greater as the weight is greater. This suggests a problem with the engine of the drill.)

Description of the Water Surface Injection in a Geothermal Plant

— Show an OHP transparency of such a plant.

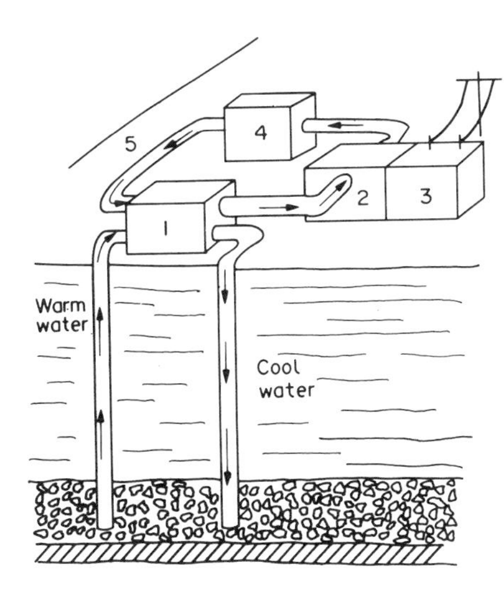

Advantages of Using Geothermal Power

Discussion with the students concerning:

(a) A natural phenomenon which is reliable.
(b) The geothermal energy cycle which is self-contained.
(c) Needing no outside support to maintain it, rail strikes, natural catastrophes, etc., would not put it out of commission.
(d) Does not involve political implications of foreign intervention.
(e) It is relatively inexpensive.

A poster or OHP transparency showing the countries which have goethermal stations. Also slides showing pictures of these stations.

In general, the capital cost of a geothermal plant is high, but the running costs are low, so that the final cost of a unit of electricity compares favourably with that from a hydro-electric source. However, there may be local considerations which affect the ultimate price.

Disadvantages of Using Geothermal Power

Discussion with students concerning:

(a) The many fractures which will need to be made in areas without naturally occurring steam or water.
(b) There have been recent earth tremors in Colorado, U.S.A., due to reinjection of water.
(c) Plants are "dirty, noisy, unsightly, malodorous and possibly dangerous" (Explosion) (*Newsweek*, 19 Feb, 1973, p. 72).
(d) Steam contains hydrogen sulphide.
(e) Other minerals present in waters can poison fish and other forms of life in streams and rivers after the steam condenses.

Other Activities

Trips to a geothermal station, if there is one in the community. The students should make a report about the station.

The class is divided into three groups. One group finds information about a dry steam geothermal plant, the second about a wet steam plant and the third about a hot dry rock plant. In a plenary session they then report their findings.

References

1. *Energy and Environment*, J. M. Fowler, McGraw Hill.
2. *Macmillan Dictionary of Energy*, Malcolm Slesser, Macmillan.
3. *Alternative Energy Sources*, R. H. Taylor, Adam Hilger.
4. *Energy, A Guide Book*, Janet Ramage, Oxford University Press.
5. *Scientific American*, Sept. 1972, pp. 70–77.

Books from the Philippines

In addition to the ideas presented above, it is worth noting that one country that has done much to develop its own geothermal resources is the Philippines and that their Institute for Science and Mathematics Education Development has produced some excellent teaching material. Much is in the form of small booklets giving information, experiments, questions and other activities. Three examples are as follows.

Geothermal Energy, An Introduction

This module provides information about geothermal energy development in a southern part of the country. The focus is on how geothermal energy is tapped and utilised to generate electricity. Theories on the sources of earth heat are briefly explained.

Electricity from Underground

This module aims to increase the students' awareness of the limited supply of energy from oil and how heat from underground can take the place of oil as a source of energy.

Electromagnetic induction, the operating principle of electric generators is demonstrated. A field trip to a nearby power station is suggested.

Geothermal Energy: its Chemical and Thermal Effects

Activities in this module include measuring pH of rainwater collected in different areas in the vicinity of the geothermal power plant and determining the effects of excess salt on growth on mongo seeds.

Worldwide Geothermal Capacity

Worldwide Geothermal Capacity (in MW) in 1982 and Plans for the Year 2000

Country	1982	2000
	(megawatts)	
United States	936	5824
Philippines	570	1225+
Italy	440	800
New Zealand	202	382+
Mexico	180	4000
Japan	175	3668+
El Salvador	95	535
Iceland	41	68+
Kenya	30	30+
Soviet Union	11	310+
Azores	3	3
Indonesia	30	92
China	4	4+
Turkey	0.5	150
Costa Rica	0	380+
Nicaragua	0	100
Ethiopia	0	50
Chile	0	15+
France	0	15+
Total	2717	17,649+

13

Teaching about Fossil Fuels

C. SHEA
Queensland, Australia

CHANGFENG JIANG
No. 4 High School, Beijing, China

AN OVERVIEW

H. Goldring
Weizmann Institute, Israel

When discussing fossil fuels (coal, oil and gas) the process of their formation should be mentioned (photosynthesis). It should be emphasised that the process of formation of these fuels took a very long time—it has been going on for 600 million years. The fossil fuels are stored solar energy, but they are being consumed far faster than they were formed. This can be compared to a father who has been saving money for his children at a constant rate and the children start spending it at a much faster rate. In the end there is nothing left.

The concept of renewable and non-renewable energy sources could be introduced at this point.

Exponential growth of fuel usage could be discussed, together with the consequent depletion of the stock of non-renewable fossil fuels.

It is important to discuss the ways in which supplies of mineable fossil fuels can be estimated. Even if the whole Earth were full of coal (or oil) and we go on using this fuel at the present rate of growth, it would not last very long. (In the case of oil, the Earth is considered as a super tanker.)

A nice experiment for students, where the density of solar radiation reaching the Earth can be measured is described in *Solar Energy—how much do we receive?* by O. Keden and U. Ganiel which is described in the section on Solar Energy.

References

1. Energy—the analogy approach, by Doris A. Simonis, *The Science Teacher,* Feb. 1982, pp. 41–44.
2. The energy resources of the Earth, by M. King Hubbert, *Scientific American,* Sept. 1971, pp. 61–70.
3. Forgotten fundamentals of the energy crisis, by Albert A. Bartlett, *Am. J. Phys.* **46** (9), Sept. 1978, pp. 876–888.

IDEAS FROM THE GAMBIA

M. H. Padakannaya
Gambia High School, Banjul, The Gambia

Aim: To develop an understanding of the nature of fossil fuels for a General Science Course.

To introduce the subject, questions might be asked about the different types of fuel used by society. Identifying the fossil fuels would lead to an understanding of the term "fossil".

Then the formation of fossil fuels could be explained and the different states in which they can exist. In the Gambia, there is no coal mining or oil drilling so *charts* showing the extraction of coal and oil would be shown. The production of natural gas would also be explained. Other charts showing the common methods of transporting these fuels could be shown and the students asked to find out which is the most economical form of transport.

The use of fossil fuels to produce heat could be demonstrated and the further production of electricity explained.

COMPARISON OF LIQUID AND SOLID FUELS—AN EXPERIMENT

C. Shea
Queensland, Australia

Aim: Students perform simple experiment to compare fuels to gain an appreciation of the characteristics of different fuels.

Strategy: Ignite and observe the combustion of a number of fuels.

Equipment: Teaspoon, small tin can lid, box of matches, range of solid liquid fuels, watch.

Procedure: For each fuel in turn:

1. Clear the lid.
2. Place on it one teaspoonful of the fuel.
3. Light the fuel with the match.
4. Observe the combustion with respect to, time taken to ignite, time taken to burn away, colour of flame, colour and type of smoke, released, nature of any residue.

5. Fill in data table.
6. Clear the lid, apply another fuel.

Fuels Data Table: Characteristics

Fuel	Ignition time	Burning time	Flame colour	Smoke	Residue	Other
Wood splinter						
Charcoal						
Kerosene						
Diesel						
Lubricating oil						
Petrol						

Note: Add any special local fuels. Warn students to be careful. Perform experiment in safe area.

Discussion: Students make comparisons of the positive and negative attributes of the fuels tested.

Extension: Students can nominate/calculate the most suitable fuel for particular situations knowing the relative local cost and availability of each fuel.

WHICH FOSSIL FUEL IS USED—AN ACTIVITY

C. Shea
Queensland, Australia

Aim: (1) To identify energy transformation in use of fossil fuel.
(2) To identify the original fossil fuel for a number of appliances and machines.

Method: (1) Students identify the energy conversions in an electricity power station, e.g.: Coal—Electricity is Chemical—Heat—Kinetic—Electrical. Select a number of others.
(2) Show students pictures of a number of appliances and machines. Students identify the original fossil fuel used in each machine.

Some pictures to show: car, iron-steel, furnace, jet aeroplane, television sets, heater, train (steam, diesel, electric), motor boat, hair drier, barbecue. Many other examples could be used.

Students complete the following table.

Appliance	Original fuel

Students should gain an appreciation of the impact of fossil fuels in their lives.

FOSSIL FUELS AND LIFE STYLE—A DISCUSSION ACTIVITY

C. Shea
Queensland, Australia

Aim: To show students that life style is dependent upon and determined by a supply of fossil fuels.

Background: Trace the historical use of fossil fuel to level of society.

Method: Compare changes in schooling over the last 100 years which have been brought about by the increased use of fossil fuels.

Consider: (1) transportation to school,
(2) fuel used for heating and lighting,
(3) writing implements (modern ball-point pens, product of petrochemical industry),
(4) type of furniture (e.g. plastic),
(5) teachers' aids,
(6) school uniforms,
(7) bitumen in school grounds.

Students can add to the list.

Additional material: Students can compare their own life style with that of students in other countries (which consume fossil fuels at different rates).

Extension: Ask students to write or discuss what would happen if fossil fuels suddenly become unavailable.

Reference

Fossil Fuels, Book 3 in "All about Energy Kit", Energy Authority of NSW, Sydney, Australia, 1983.

ENERGY AWARENESS—AN ACTIVITY

C. Shea
Queensland, Australia

Aim: To show students that energy matters, particularly those relating to fossil fuels which are important socially and politically.

Method: Over a period of 6 weeks, students collect newspaper clippings on all energy matters, e.g. oil finds, coal miners' strikes, new power stations, etc.—anything of economic, social or political significance. The file could be divided into sections—local and overseas.

A map of the student's own country and a world map is drawn. Each item in the file is coded and located (local or overseas) on the map.

A legend needs to be constructed to assist placement on the map.

ENERGY USERS' INVENTORY: A DATA ANALYSIS ACTIVITY

C. Shea
Queensland, Australia

Aim: To make students aware of their own use and dependence upon fossil fuels.

Methods: Students write down all energy appliances in their own home. For each appliance, students write down a substitute for each appliance. Students calculate the number of kilowatt hours (or any other convenient units of energy, perhaps joules) each appliance consumes annually, as well as the cost of running each appliance.

Students construct and complete a table like this:

Energy Inventory

Appliance	Substitute	Kilowatt hours used annually	Annual cost

Reference

Energy, Victoria Department of Education, Melbourne, Australia, 1982.

FOSSIL FUELS AND THE FUTURE: A DISCUSSION ACTIVITY

C. Shea
Queensland, Australia

Aim: To stress upon students that fossil fuels are non-renewable resources.

Method: Burn some oil in the classroom. Ask students if the oil could be used again. Define "non-renewable" resources. List all fossil/non-renewable fuels (students should include electricity, if appropriate in this list).

Further Questions: Why was there an energy crisis in 1974 and 1980?

Is there an energy crisis for fossil fuels today? (Able students will be able to calculate rates of supply, use and amount of reserves.)

Discuss exponential growth. Students debate/decide the need for conservation of fossil fuels.

Students outline possible conservation strategies for their country and themselves as individuals.

CHANGING WORLD FUEL CONSUMPTION

Changfeng Jiang
No. 4 High School, Beijing, China

Aim: To put our present fuel consumption into an historical context.

Discuss the diagram below which shows the changing pattern of the consumption of the major sources of energy up to the present and with projections into the future. The use of a logarithmic scale is worth noting.

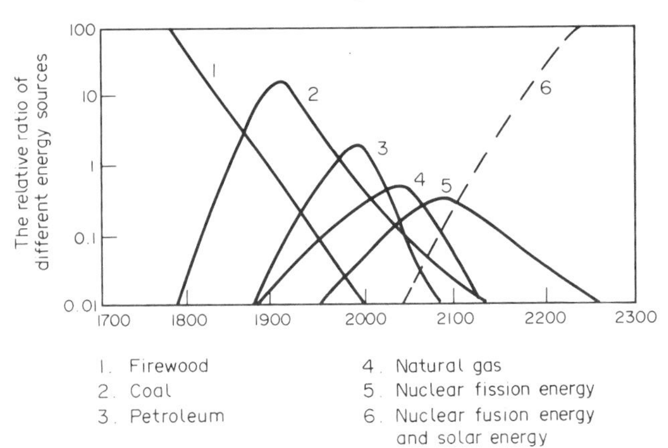

FIG. 1. Changing Patterns of Energy Consumption in World History

QUESTIONS ABOUT COAL—DISCUSSION QUESTIONS

Changfeng Jiang
No. 4 High School, Beijing, China

Aim: To enhance awareness of the advantages and limitations of one fossil fuel.

1. How does one nation's changing use of coal compare with that of the world?
2. What is the world (or national) distribution of sources of coal?
3. Does the world (or national) distribution consumption of coal match the distribution of sources?
4. What new technologies are being developed to extend the use of coal? (Such technologies would include liquifaction and garification.)
5. Comparing coal with oil, what are its advantages and its disadvantages? (Some factors which might be considered are: size of reserves, ease of extraction and use, use as a chemical feedstock, calorific value, efficiency of combustion, pollution, ease of transport, use as a vehicle fuel.)

THE POWER STATION PROJECT—A DECISION-MAKING GAME

T. D. R. Hickson
King's School, Worcester, U.K.

Aim: To help students appreciate the complexity of real decision-making: that economic and social factors are just as important as technical considerations:

A region in a country is imagined to require a power station. Having learnt about the production of electricity from coal, oil and nuclear fuel, the students are divided into three groups. Each is given the task of examining the feasibility of building a power station using one of the fuels. They also have to choose one of six possible sites for their power station and in so doing, take into account technical, economic, social and environmental factors.

Then each group puts forward a detailed case for its particular power station. After a discussion of the relative merits of the three schemes one is chosen.

This particular exercise is produced as a package containing a teachers' guide and ten students' books for the Science in Society Project by the Association for Science Education, College Lane, Hatfield, Herts. AL10 9AA, U.K. However the idea could be used in most national settings.

FOSSIL FUEL AWARENESS KIT

C. Shea
Queensland, Australia

Aim: Students prepare a display (e.g. posters, slide/tape, video) and perhaps a brief lecture on the finite nature of fossil fuels and the need for conservation in society generally and on a personal level.

Method: Students in the class collect data about fossil-fuel depletion.

Students assemble the appropriate data and make posters and possibly other audiovisual material together with a script.

The "public awareness" kit is assembled to give the public at large a convincing case for the need to conserve fossil fuels and to reduce their use by citizens.

The "public awareness" kit can be displayed within the school, within other schools, and in centres outside the school; for example in shopping centres, banks and in public buildings.

Note: In preparation and presentation of the material, students themselves have a very potent learning experience.

14

Teaching about Nuclear Energy

A. B. PRAT AND OTHERS

Torino, Italy

This programme is intended to illustrate activities some of which will be more appropriate than others according to the facilities available in a particular school.

In many cases, of course, it would be sensible to adapt the techniques mentioned to fit the students' own locality. The individuals named in what follows can, of course, be contacted for further details.

How to Introduce the Subject

As always, when introducing a new (and in this case, difficult) subject it is important both to stimulate the pupils' interest and to provide a general overview which will give meaning to the more detailed study to follow. Some possibilities are as follows.

(a) Get the pupils to collect newspaper articles on nuclear power and then to extract from them the main issues.

(b) Show a film or a set of slides.

(c) Give an information sheet with a set of contrasting quotations from authorities, books, newspapers, etc., and ask pupils to extract from them some key questions.

(d) (i) Arrange a discussion between experts. (This activity is most effective if used after the relevant information and concepts have already been introduced.) Where it is not possible to obtain a group of experts, audio cassettes of such discussions are often available. Furthermore, discussions of this type are to be found in science magazines.

(ii) As an alternative to the above, the students themselves can be the participants. In this case, where the students are role-playing, there is a range of forms which the activity may take. At one extreme, all the information needed may be given to the students, even to the words they speak. Vivien Talisayon (Philippines) provided one good example of such material, "Nuclear Power: Pros and Cons" prepared by the University of the Philippines Institute for Science and Mathematics Education Development. Here views and data are given, for and against, the first Philippine Nuclear Power Plant at Morong on the Bataan penninsular. The discussion deals with the following topics.

1. Environmental and Health Impact.
2. Safety of Plant Design.
3. Characteristics of Plant Location.
4. Waste Control and Management.
5. Economic and Political Implications.

At the other extreme, as Chris Shea (Australia) reported, students can be assigned a role to act in a debate and then told to prepare their own cases. This, of course, provides an opportunity for students to develop a range of skills but it is time-consuming: teachers will need to acquire material from as wide a range of sources as possible and will need to provide general background information; students will need several lessons to prepare their cases, especially to digest the technical details.

The actual debate will need to be conducted formally and preferably adjudicated by someone other than a student or the teacher. Such debates tend to lead to further discussion amongst the students but now it should be informed discussion.

The Physics of Fission

For some students, it will be appropriate that they learn something about the process of fission; others may simply need to know that it can occur.

To avoid asserting, without any evidence, that nuclei are formed from protons and neutrons, students could deduce this fact from a table of atomic numbers and mass numbers for the first ten elements in the Periodic Table.

Symbol	H	He	Li	Be	...	Ne
Atomic number Z	1	2	3	4		10
Mass number A	1.008	4.00	6.94	9.01	...	20.18

It is not necessary to give details about the unit used for atomic mass (it could be said to be a conventional unit that may be defined later). After making it clear that, in a neutral atom, Z is also the number of electrons, the following questions can be asked.

(a) Could these nuclei be formed by A protons only?
(b) Could they be made from A protons and $(A–Z)$ electrons?
(c) Could they be made from Z protons and $(A–Z)$ other particles? What properties would these other particles have to have assuming they were all the same?

In the discussions that follow, possibility (a) will obviously be discarded because atoms could not be neutral, (b) and (c) are of course both apparently acceptable. For (c) to work, the other particle must be electrically neutral and of about the same mass as a proton. That such a particle—a neutron—was eventually discovered could then be revealed.

Obviously the values of A can only be explained approximately (for this reason it is best to consider only the first element where there are few isotopes) and this opens the way to introduce isotopy.

Having established that nuclei are made of protons and neutrons, the question which must arise is, how is the nucleus held together. It is fun for those who can to calculate the electrical force of repulsion between two protons "in contact". If we imagine their centres to be separated by, say, 2×10^{-15}m, the force will be almost enough to raise a 6 kg mass off a table!

Since nuclei do hold together then there must be some special force which will hold proton to proton and perhaps proton to neutron. These forces must be huge and must have a very short range since we know that, positively charged particles can be fired at nuclei so fiercely that they can get very close and yet they are still repelled.

Thus, in a simple picture, if two parts of a nucleus separate just enough for the electrical force of repulsion to be greater than the nuclear binding force then they will fly apart at very high speeds. It is this kinetic energy which is used in a nuclear reaction.

It turns out that such a process of fission can occur spontaneously, but very rarely, for some of the heavy elements including an isotope of Uranium U-235. However, if a neutron hits a U-235 nucleus then that nucleus is very likely to break in two. Furthermore, as well as breaking into two roughly equal parts, the nucleus emits two or three neutrons. These can then induce further fissions.

When dealing with the energy released during fission, the conference was reminded of the need to avoid the trap into which many text book writers have fallen. Hanna Goldring (Israel) contributed the following.

A point that is worth stressing in this context is that in some books an approach which may lead to misconceptions is presented. The reason that in exothermic nuclear reactions so much energy is released is said to be Einstein's theory of relativity. In many books you find statements like ". . . the source of nuclear energy is not in electrical or gravitational forces . . . the basic explanation of the energy released in nuclear reactions lies in Einstein's theory of relativity, according to which the relationship between mass m and energy E

is $E = mc^2$, and the reason that such large quantities of energy are released is that c is so large".

Actually, Einstein's formula implies that energy and mass are equivalent, and the relationship between mass and energy is $E = mc^2$. Instead of the two conservation laws of mass and energy, we now have one, the law of conservation of mass-energy.

Every time a change in energy occurs, a change in mass also occurs. For example, when we heat a given quantity of gas the average kinetic energy of its molecules increases, and therefore its mass also increases. The amount of increase of the mass in this case is however so small, that it cannot be measured. In many chemical reactions energy is released. In these reactions the energy released is of the order of a few electron-volts per reaction. In this case too a change in mass occurs, but again it is too small to be measured. For example, when 1 kg of gasoline is burnt 5×10^7J of energy is released (the heat of combustion of gasoline). This energy is equivalent to a mass $\frac{5 \times 10^7}{9 \times 10^{16}}$ kg which is 5.5×10^{-10} kg. The decrease in mass of 1 kg of gasoline is 5.5×10^{-10} kg. It is extremely difficult to measure such a change in the mass of 1 kg. But when 1 kg of uranium undergoes fission 8×10^{13}J is released, which is equivalent to a mass of 9×10^{-4} kg. This can be measured rather easily.

Just as nobody would claim that the source of chemical energy lies in the equation $E = mc^2$, we should not say that the source of nuclear energy lies in this formula.

The energy released in chemical reactions is derived from the electrical forces between the atoms, and the energy released in nuclear reactions is derived from the nuclear forces between the nucleons.

The difference between nuclear and chemical reactions lies first and foremost in the size of binding energies, which comes from the difference in the strength of electrical and nuclear forces.

To illustrate the process of a chain reaction, Vivien Talisayon (Philippines) proposed a simple model in which bottle caps are taped in pairs to represent nuclei. The pairs are connected by matchsticks so that when the first matchstick has been lit it causes the first pair of bottle caps to separate and the next matches are set on fire.

Since the control of chain reactions depend on the interplay of several parameters it is not easy for young students to understand. Thus Hanna Goldring (Israel) proposed a simple visual scheme that stresses the importance of the neutron cycle.

Control of the Chain Reaction in a Reactor

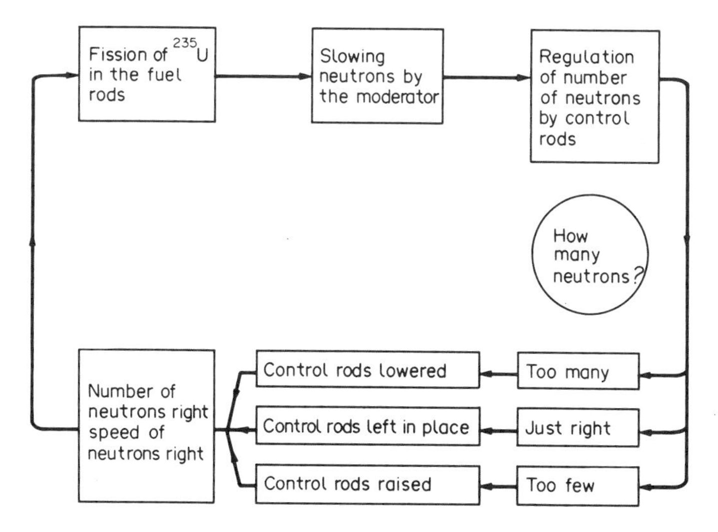

FIG. 1.

This, of course, introduces the use of a moderator and the need to slow the neutrons.

If a computer or a programmable pocket calculator is available, students can "play" with the parameters which represent the qualities of the moderator, of the absorber/control rods and of the fuel.

The model is illustrated by the following diagram (Fig. 2).

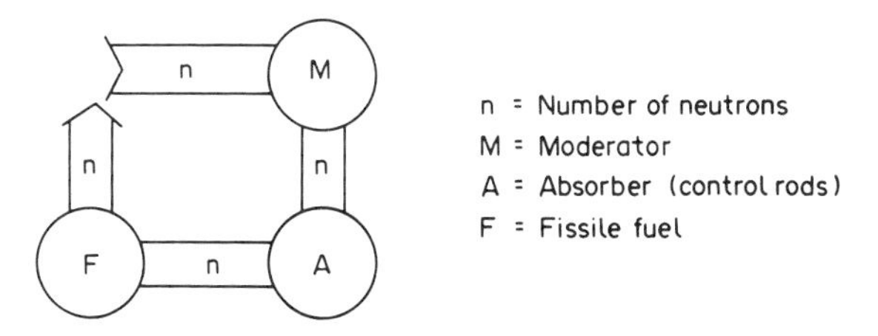

FIG. 2.

(a) The incoming neutrons are slowed down by the moderator. In this phase there is a probability M that they are absorbed by the moderator itself, or that they escape from the reactor core. M depends mainly on the type of moderator (in principle it should be as low as possible).

(b) Neutrons are also absorbed by the control rods, if they are not completely raised, with a probability A.

(c) Finally, neutrons can cause nuclear fission with a probability F (related to the quality of the fuel, mainly on how it is enriched). Each neutron that causes a fission generates, on the average, three neutrons. Pupils can vary A—thus "moving" the control rods. If A is too high, the number of neutrons will fall rapidly; if it is too low it will rise exponentially. Adjusting M to the right value, it will fluctuate but remain approximately constant, it is also possible to vary F ("enrich" the fuel) and correspondingly vary M (use natural water with enriched uranium, instead of heavy water with natural uranium, etc.).

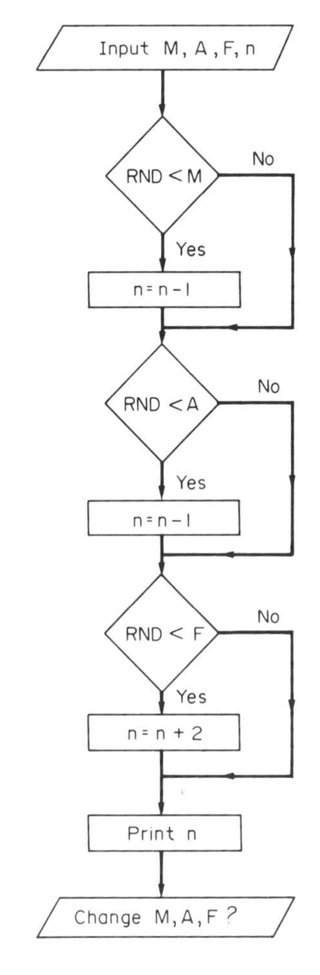

FIG. 3.

(There are, of course, an increasing number of commercial computer programs which simulate the operation of a nuclear reactor, its turbines and generators.)

Technology

It is important that pupils be aware that a fully developed industrial plant is much more complicated than the scientific concepts on which it is based.

A Visit to a Nuclear Power Plant

This is of course possible only if a nuclear power plant is near to the school site. If this is not the case, many problems can also be illustrated on a poster, but it has not the same impact. Useful help can come from a visiting expert, i.e. a person that has worked with nuclear plants and can communicate personal experiences. Since a visit to a nuclear plant requires time, organisation, and money, it must be organised carefully. One suggestion is to have a discussion period in which pupils prepare a list of the relevant features they want to observe and the questions they want to pose.

Here are some examples:

- What is the working environment in a nuclear power plant? Are people worried about their health? How many people work in it?
- What are safety measures in the power plant? Hint: Try to pass near a radiation detector with a fluorescent watch.
- A nuclear power plant is made of conventional parts (electricity generator) and a nuclear one. Which one is more likely to have failures and requires more repairs and maintenance?
- What is done with waste heat? Are there signs of environmental pollution?
- What is done with radioactive wastes? Ask to see the storage facilities and get an idea of the quantity of wastes.
- What are the main problems in the daily life of the plant?
- How is the life of people near the plant affected by it?
- How long did it take to be built?

Some questions which might be considered are:

- Discuss advantages and disadvantages of various types of conventional nuclear plants (natural water, heavy water, gas graphite and possibly others).
- Give pupils a list of elements and compounds and ask them to decide whether they would make good moderators. Here is a possible list: iron, gold, helium, lithium, fluorine.
- Evaluate years of availability of nuclear fuel, at present consumption rate:

  (a) if only conventional reactors are used,
  (b) if breeders are used.

What if rate of consumption was to grow at 5 per cent per year? (Pupils may answer this question by using a pocket calculator, multiplying by 1.05 each year, and seeing when they get above the reserve threshold—No need to use logarithms or exponentials.)

General Issues

Like any other power plant, fission reactors have advantages and disadvantages. Many are common to other devices (environmental pollution, possibility of accidents, economic features, etc.), some are peculiar of this energy source. These are mainly connected with radiation, so a course on nuclear energy may presumably involve some activities on radiation.

Experiments on Radiation

(a) Small, inexpensive and easy to build cloud chambers operating with a low intensity source are widely available.
(b) Radioactive materials, such as uranium ore, that can be available in the Earth Science Department, can leave a trace on a photographic film. (See *Project Physics Course Handbook*, Unit 6, Holt and Rinehart, U.S.A. This method reproduces the experience by which Becquerel discovered radioactivity at the end of the nineteenth century.)
(c) Another possible radiation detector is a sensitive electroscope: brought near a radioactive source it will discharge due to air ionisation.

Pencil and Paper Activities on Radioactivity

(a) List radioactive sources in the environment and work out the total amount of radiation received every day. *(Thinking about Energy—Elementary studies guide,* State of Delaware, U.S.A. produced the example opposite. By filling out this form, students get an idea of the amount they are exposed to every year.)
(b) Ra^{226} decays spontaneously through the following decay chain:

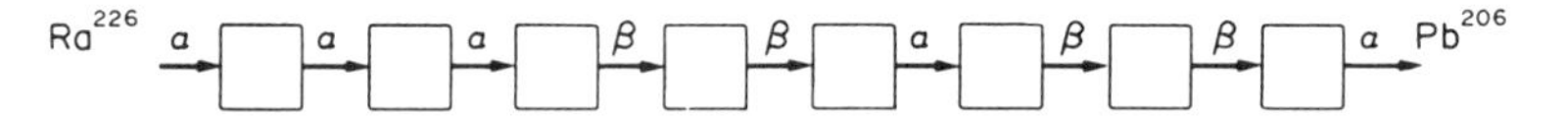

Using an isotope chart, fill in the blanks with the name of the corresponding element and its mass number (other decay chains can be used).
(c) List the precautions taken in a nuclear power plant against leakage of radiation in the environment,

(i) during normal functioning,
(ii) in case of accidents.

Do you think some of them ought to be enhanced? How? (This question is very interesting if used in preparing a visit to a nuclear reactor.)

	Common Source of Radiation	Your Annual Inventory
WHERE YOU LIVE	Location: Cosmic radiation at sea level Add 1 for every 100 feet of elevation 1 × ________ Typical elevations: Pittsburgh 1200; Minneapolis 815; Atlanta 1050; Sequoia National Park 6400; Las Vegas 2000; Denver 5280; St. Louis 455; Yosemite Valley 4000; Salt Lake City 4400; Dallas 435; Bangor 20; Sacramento 30; Spokane 1890. (Coastal cities are assumed to be zero, or seal level.)	40 —
	House construction: Wood 35; Concrete 50; Brick 75; Stone 70	________
	Ground: U.S. Average	15
WHAT YOU EAT, DRINK & BREATHE	Water, Food, and Air: U.S. Average	30
HOW YOU LIVE	Jet Airplanes; Number of 6000-mile flights ________ × 4	________
	Television viewing: Black and white—Number of hours per day ________ × 1 Colour —Number of hours per day ________ × 2	________ ________
	X-ray diagnosis and treatment: Chest X-ray ________ × 100–200 Gastrointestinal tract X-ray ________ × 2000 Dental X-ray ________ × 20	________ ________ ________
	Compare your dose to the *U.S. Average of 200*	Sub Total ________ mrem
HOW CLOSE YOU LIVE TO A NUCLEAR PLANT	At site boundary: Number of hours per day ________ × 0.2 One mile away: Number of hours per day ________ × 0.02 Five miles away: Number of hours per day ________ × 0.002 Over 5 miles away: ________ None	________ ________ ________

Total ________ mrem

One mrem per year is equal to: Moving to an elevation 100 feet higher
Increasing your diet by 4%
Watching one additional hour of black and white TV per day
Taking a 4–5 day vacation in the Sierra Nevada Mountains

FIG. 4.

Risk Analysis

The technique for analysing risks is interesting in itself and can be applied to a variety of situations. A way of introducing it is to apply it to a familiar situation. A person is injured (broken leg) in a car accident. What is the probability of complete recovery? We can break down the chain of events following the accident in the following way:

(a) Transportation to a hospital. There is a slight probability. (Let's suppose 1 per cent that a bad journey damages the leg in an unremediable way.)
(b) In the hospital, the doctor that takes care of the injured leg may or may not act correctly. The probability will depend on the kind of injury, the competency of the doctor, etc. We estimate 90 per cent as the probability of correct medical care.
(c) Complications that prevent healing may or may not arise: We take that complications have 10 per cent probability in the case of correct medical care, 50 per cent in the contrary. We have the following "risk tree".

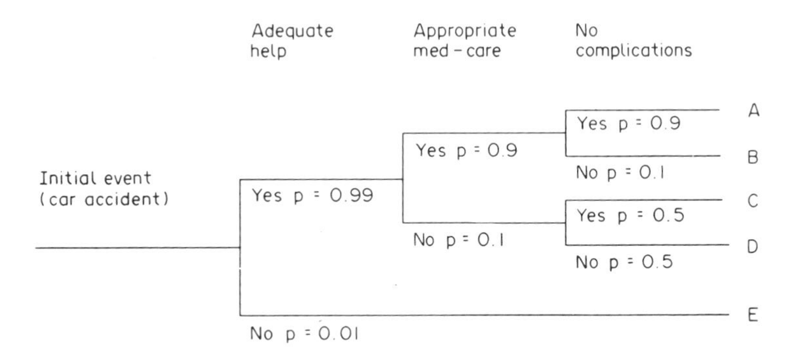

FIG. 5.

Complete healing corresponds to cases A and C. Its probability may be obtained calculating the probability of the two branches and then summing them.

$$P(\mathrm{A}) = 0.99 \times 0.9 \times 0.9 \sim 0.80$$
$$P(\mathrm{C}) = 0.99 \times 0.1 \times 0.5 \sim 0.05$$
$$P = (\text{healing}) \sim 0.85$$

Similar examples may be invented by students (probabilities of winning a tennis cup or football championship). This is exactly the technique used to evaluate risks in nuclear reactors. Students may discuss advantages and disadvantages of such a technique. (Described in A. Bastai Prat, *Fission Nucleare,* Zanichelli, Italy, 1984.)

Discussion

At this stage it is possible to have a general discussion on the problem of nuclear reactors. B. G. Kusuma (India) described one such discussion.

Plan of the discussion

1. *The topic:* Nuclear energy: What it is? What are its uses and abuses?
2. *Duration:* 15 minutes.
3. *Participants:* IX Standard students.
4. *Operation:* The class was divided into groups. Group leaders were selected. The pupils were given three days in which to gather information. The group leaders spoke in the discussion, whereas others were active listeners. The discussion was tape recorded.

The class participated enthusiastically in this discussion. A sample of questions raised during the discussion is given below:

1. Do you know a natural source of nuclear energy?
2. How can man generate nuclear energy?
3. Where are the nuclear reactor plants in India?
4. What happens if the nuclear energy production in a reactor is not controlled?
5. What are the harms that a nuclear reactor may possibly cause?

Many more questions were raised and answered satisfactorily. The pupils were able to distinguish between fusion and fission.

The outcome of this strategy to teach this particular topic can be summarised as follows:

1. It instilled an interest in learning.
2. It inspired the pupils to collect facts and information from different books and magazines.
3. It kindled an awareness of national and international problems relating to the energy possibilities.

Flow diagram of the Strategy

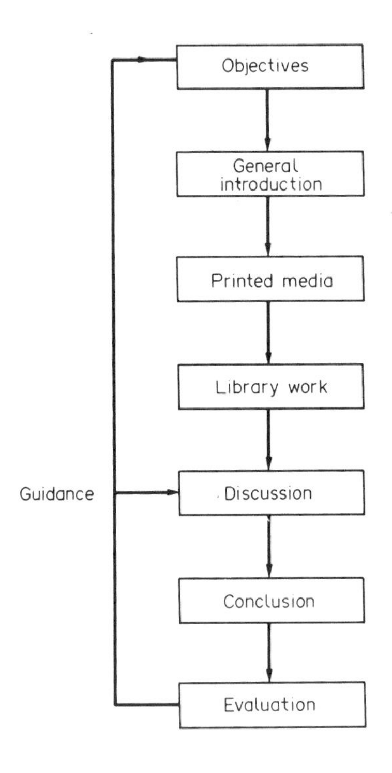

FIG. 6.

Simulation Games

An effective tool for learning about nuclear energy are simulation games. One such game is the Power Station Project, which is part of the Science-in-Society project and obtainable from the Association for Science Education, College Lane, Hatfield, Herts AL10 9AA, U.K. In this, teams have to decide whether to build a coal-fired, an oil-fired or a nuclear power station.

Two other simulation games of interest are published by the NSTA in *Playing with Energy* (Washington, U.S.A., 1981). "Consultants Inc" is concerned with electricity generation and its economics. "Conmexus" is concerned with the energy problem related to the energy resources available in three countries—Canada, U.S.A. and Mexico.

Nuclear Fusion

Nuclear fusion was not considered in Bangalore as the preoccupation was mainly with existing technologies. However, one interesting approach has been contributed by Roger Humphreys (U.K.) as follows.

It has been suggested that the following process might be harnessed to provide useful energy on Earth:

$$^{2}H + ^{2}H \rightarrow ^{3}He + ^{1}n + 3.3 \text{ MeV}$$

How many hydrogen atoms are there in 1 m^3 of water?
How many of these atoms are ^{2}H? (1 in 7000).
How much energy would be available from the fusion of two ^{2}H nuclei?
How much energy would be available from 1 m^3 of water?

The oceans are a vast source of water.
Estimate the volume of the oceans. (1.5 × 1018 m^3)
How much energy would be available from this volume by the fusion process discussed?
How does this amount of energy compare with the annual world consumption of energy?

What factors would limit the use of this fusion process in providing energy on a large scale?

15

Teaching about Biomass

B. G. KUSUMA

Acharya Pathasala Girls High School, Bangalore, India

T. P. SUKUMARAN

Kerala, India

Naturally, most of the activities described below use materials particularly appropriate to Southern India. However, teachers in other parts of the world will often be able to use local materials although whether their climatic conditions will be helpful is another matter!

What is Biomass?

(a) After a discussion, list all the fuels which are burnt to produce heat. Concentrate on those fuels which are available locally.
(b) Get pupils to bring twigs, dried leaves, paddy husk, coconut fibre, shells and dried cow dung. In as near controlled conditions as possible, burn these fuels so that pupils can observe and compare. Which make the most effective fuels should be discussed.

How Can More Energy be Obtained from Biomass?

(a) Small groups of students should bring some wet cow dung in a tin can. They should mix it with the same quantity of water, stir well and then close the lid. After 24 hours have elapsed, they should remove the lid and they will smell and perhaps recognise the gas that is being produced. Stirring will reveal bubbles.
(b) If instead the can is sealed with a lid fitted with a tube at the end of which is a used hypodermic needle, then the gas can be burnt.

The gas will be produced for many hours so that if the can is to be left sealed, say overnight, it should be kept in cold water to slow down the rate at which the gas is produced. Not only will this prevent the pressure building up to a dangerous level but it emphasises the important point that the reaction goes faster at higher temperatures.

For those pupils for whom it is appropriate this could be a useful opportunity to discuss anaerobic fermentation as a process which has a wide importance.

A Field Trip

If it is possible to visit a local biogas production centre, this is a very rewarding activity. For maximum benefit, pupils must be carefully prepared for such a visit. Then their questions and their observations will produce much useful information for later activities.

If a new production centre has been set up it is useful to return several times to monitor changing patterns of use and production and to see whether problems have occurred.

Debates and Discussions

These need to be carefully prepared. Subjects might include:

(a) "Should cow dung be used as a fuel or a fertiliser." (This should bring out the point that if cow dung is used to produce biogas, the gas burns twice as efficiently as dung cakes *and* the residue sludge is almost three times better as a fertiliser.)
(b) "What are the health consequences of using dung to produce biogas?"
(c) "Which is best, a community biogas plant or an individual plant?" (This should bring out some interesting economic, political and social points.)
(d) "What are the advantages and disadvantages of using biogas as a fuel?"
(e) "What other than cow dung could be used to produce biogas." (It is important here to consider both plant wastes and human wastes. The latter could be considered in (b).)

Experimental Investigation

What would be the effect of varying the proportion of water to dung?

What Happens in Other Countries?

Here pupils can collect newspaper and magazine articles, refer to books and even correspond with schools and colleges in other parts of the world.

Tape and Slides

When discussing, say, the research being done on the use of wood chips to produce biogas, a field trip may be impossible. Even if it could be arranged, it might well be ineffective as access for a whole class may be limited and the density of technical information might be too great. In this case, with the help of the researchers, a teacher could make a tape recording and collect slides so as to present the material at just the right level and pace for his or her pupils.

At the beginning of this section, the use of wood and other materials, as a fuel to be burnt, was briefly mentioned. In fact, of course, the use of such fuel for cooking is its major use in the world today.

Don Kirwan (U.S.A.) has provided the following thought-provoking question.

Question: Can wood serve as the energy source which will satisfy the energy need of the U.S.A.?

The U.S.A. has a land area of 9.36×10^6 km^2.
Wood has been shown to have an effective yield under optimum conditions of 16 metric tons/acre. According to McElvoy in "Utilization of land with limited capabilities", a paper presented at the Conference on Biomass in Kansas City, U.S.A. in 1977. There are available for use a total of 100×10^6 acres of unused cropland.

Thus:

$$(16 \text{ metric tons/acre}) \times (2200 \text{ pounds/metric ton}) \times (2266 \times 10^6 \text{ acres}) \times (30.8 \times 10^6 \text{ BTU}/4400 \text{ pounds}) \times (1/3413 \text{ BTU/kWh}) = 163 \times 10^{12} \text{ kWh}.$$

We will assume that the U.S.A. needs 1.0×10^{13} kWh of energy.

Consequently, we need to have only 1/16.3 of the total land area of the U.S.A. for wood. This means that we require 139×10^6 acres, but there are only 100×10^6 acres available. The answer must be NO!

One of the most serious consequences of the world's need for fuel wood is, of course, deforestation. The World Wild Life Fund, together with the International Union for the Conservation of Nature, have produced a tape/slide package called "Vanishing Forests". This has already proved to be a superb means of introducing this problem and its possible solutions to a wide range of audiences.

Finally Vivien Talisayon showed another of the excellent booklets produced by the Institute for Science and Mathematics Education Development of the University of the Philippines. It too contains information, activities, questions and experiments to be done with simple home-made equipment. This is called "Waste Not! Want Not!" It describes a method of converting waste (animal manure) to biogas and goes on to compare methane production from different ordinary waste materials. Instructions are given for building a family-sized biogas generator.

16

Strategies for Teaching Conservation of Energy

J. DUNIN-BORKOWSKI

University of Warsaw, Poland

M. H. PADAKANNAYA

The Gambia, Gambia High School, Banjul

Concept of the Conservation of Energy

"The total energy of the Universe is a constant" gives an idea that energy does not get lost or reduced, but only gets changed from one form to another. If the total energy is a constant, why should energy be conserved at all? By "conservation" what we actually mean is the saving of energy reserves for future use. Energy reserves include all the primary forms of energy which start the chains of energy transformation taking place in Nature. Degradation of energy takes place during this process.

Activity 1

Each member of the class is asked to make a list of energy uses in his or her house and estimate the amount of different kinds of energy used. An example taken from *Science in Society, Teachers' guide*, Unit F, Heinemann and ASE, U.K., is shown below.

Fuel	*Amount*	*Unit*	*J*
Electricity	4000	kWh	1.4×10^{10}
Natural gas	1500	Therm	15.9×10^{10}
Coal	1	tonne	2.8×10^{10}
Wood	0.1	tonne	0.1×10^{10}
Petrol	2000	litre	8×10^{10}
Food	4,000,000	kilocalorie	1.7×10^{10}
		Total	29.9×10^{10}

Activity 2

(a) Estimate the cost of energy consumption in your family budget.
(b) What percentage of the total family expense is spent on energy consumption?
(c) How can we cut down the unnecessary use of energy?

Activities 1 and 2 would help in gathering information on the need for saving energy.

The teacher should provide information about energy consumption of the locality, of the town and the country to make the students aware of how the expenditure of energy affects the economy of the country. Perhaps giving information about expenditure on different areas including energy consumption and asking the students to draw a pie-chart would help. More information on energy consumption per person in different countries compared with the student's own country can be given.

There are three types of strategies for conserving energy:

1. *Curtailment* in which we cut down the use of energy to the barest minimum need. This method costs nothing.
2. Improving the *device efficiency* in which later models are devised in such a way that its efficiency improves. For example, devising a car which consumes less fuel.
3. Improving the *system efficiency* where you encourage a society to change its living habits, to get the maximum benefit out of the minimum energy reserve consumed.

These three types of strategies for conservation of energy can be introduced by further classroom activities.

Activity 3

Classify the following habits as good habits and bad habits.

1. Keeping the lights on when you leave the room.
2. Keeping the bedroom windows open while sleeping during the summer.
3. Keeping the bedroom windows closed and switching on the fan during the summer.
4. Having a short hot-water shower instead of a bath.
5. Taking a bus for a long journey instead of driving a car.

Activity 4

Make a list of good habits and regularly check your performance.

An example taken from *All About Energy* kit, Book 6, *Conservation and Alternatives,* Energy Authority of New South Wales, Australia, is shown below.

Here is a list of things that you might do to save energy at home. Can you think of any more? Add them to the list below and complete it every night for a week. Tick the box when you save energy.

Energy	*Mon*	*Tues*	*Wed*	*Thurs*	*Fri*	*Sat*	*Sun*
1. Turn off the light							
2. Turn down the fan							
3. Walk to the shops							
4. Turn off the hot water tap to stop drips							
5. Close the fridge door quickly							

Activity 5

A class experiment on heat transfer and insulating materials.

Fit a box with a heater at one end and a thermometer at the other. An electric light bulb can act as a heater. The heater is switched on and the item taken for the temperature to rise by 1 °C is noted. Similar sheets of different materials like wood, metal, cardboard, etc., are kept in between the thermometer and the source of heat and the times needed for the temperature to rise by 1 °C are noted in each case.

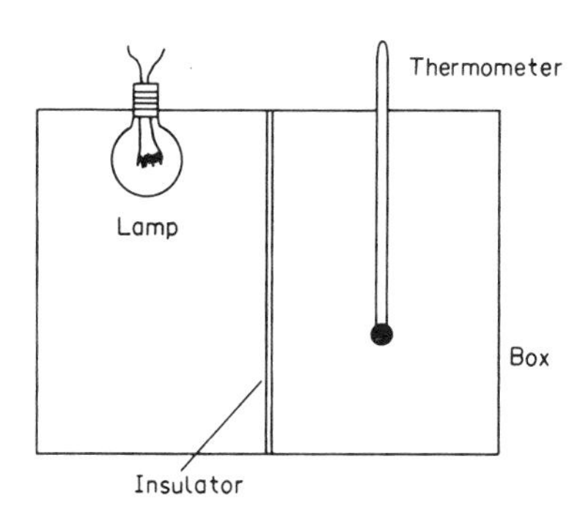

From the results the students should be asked to deduce which material would make the best insulator.

Follow-up Work

1. Estimate the heat loss for a typical house in your locality and suggest methods by which the energy waste can be cut down using insulation. Say what materials are being used and how they are being used.
2. Compare the efficiency of cooking in the open fire used in some villages in India and the wood stoves used in some other homes.

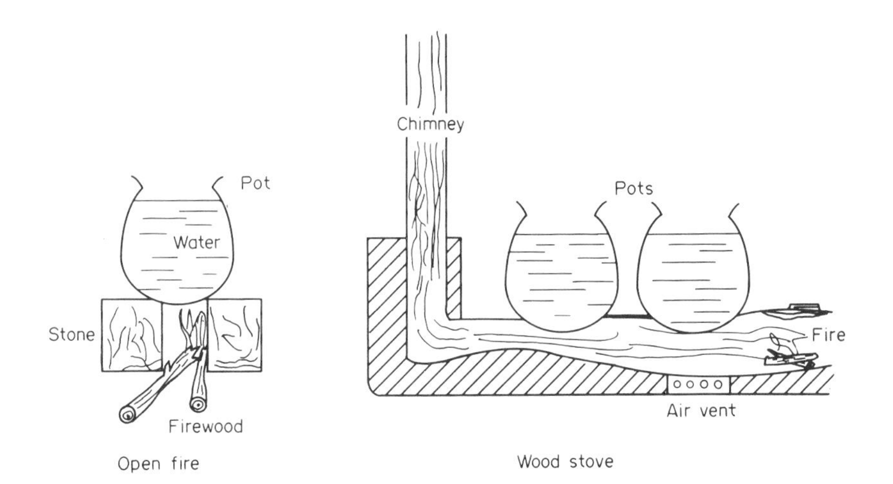

Activity 6

Collect data from the class about patterns of individual and social consumption.

1. A good example to examine might be *transport*.
   (a) What is the expenditure on transport by those students who travel by individual means such as cars, bicycles, etc.?
   (b) What is the expenditure on transport by those students who use public transport, such as trains, buses, etc.?
   (c) Discuss the energy consumption for different modes of transport.
2. Make a survey of any other energy supply used by the community as a whole such as water supply, supply of electricity, supply of gas, etc. Do you think that energy consumption is involved in the production of other goods such as implements used in a forge which serves the whole locality?

Follow-up Work

Teacher should give data about the national distribution of energy consumption in the country. Then the following questions might be used to promote discussion.

1. Which sector of the economy consumes the most energy?
2. Which sector is the next largest?
3. How does the energy used for industry compare with that needed for domestic purposes?
4. In which sectors are there possibilities for saving energy?
5. How do changes in energy consumption effect the living conditions?

Activities 3 and 4 would help in explaining how cutting down the use of energy conserves the energy reserves.

Activity 5 is to explain how device efficiency helps in the conservation of energy.

Activity 6 is to make the students understand that improvement in the system efficiency is a method of saving energy, therefore changing the living habits of society to the mutual benefit of the individuals.

Notes

1. There are posters available depicting the economical use of energy which can be shown as follow-up work for Activity 5. For example, one might show different models of car, with improvements in efficiency, along with data on their performance.
2. Visits to the places where energy-saving appliances are used can be arranged, such as thatched houses, wood-stoves, etc.
3. Tapes and audio cassettes can be played, films and videos can be shown on energy saving in different countries to depict the techniques used in conservation of energy elsewhere.
4. Debates could be arranged in order to impart the spirit of healthy competition which could finally lead to improvement in system efficiency.
5. Computers can be used, where available, to compute and simulate the heat losses during the chains of transformation.
6. Decision-making games such as choosing the correct type of stove for a particular society after considering the advantages and disadvantages of different types of stoves in use.
7. Articles in newspapers and magazines appear frequently, dealing with this topic. Students should be encouraged to collect these.

SECTION D

Energy Education at the Tertiary Level

Introduction

The first article in this section deals with general teaching/instructional strategy in a university level energy course. Specific suggestions are given which have been proven effective in energy education at this level. The remainder of the section on tertiary education is given by the tertiary group at Bangalore. They opted to construct two university syllabi for use by interested faculty. They focused primarily upon the needs of local colleges and universities in underdeveloped countries. One syllabus outlines the course content for non-science/engineering students and the other specifically addresses the science majors' and the pre-engineers' needs.

17

The Teaching of Energy at the Tertiary Level

J. DUNIN-BORKOWSKI

University of Warsaw, Poland

Since the energy crisis, issues on or related to energy are frequently discussed and widely debated. Articles on energy appear often in the columns of the local newspapers, magazines, and popular journals. It is the current social issue having scientific-technological implications and therefore is relevant to the education of all students. The teaching of energy has become more important both in the sciences and in other related disciplines. The topics on energy, or more correctly the concepts or principles related to energy, are being taught under various headings (or courses) at our local universities. Some examples are: Energy Systems, Solar Energy, Energy and the Alternatives, and Energy for Agriculture, besides the traditional topics on energy in physics, chemistry and biology. The content of these courses is inter-disciplinary and covers social, economic, and political principles and the significance of energy, its basic principles, sources, alternatives, and technologies, and its relationship to industry and agriculture.

Method of Presentation

In determining the method of presentation, it may seem tempting to use common procedural methods of science and scientific processes, which are characterised by critical thinking and based on relevant facts which lead to a logical conclusion that can be carefully tested. However, more often than not, like most courses at the tertiary level, this is not the case and the common mode of presentation is by lecture. This method is suitable when the number of students is large and the courses aim is to present an overview and an acquaintance with a range of materials and issues. However, to prevent the danger of the course becoming just more information to be learned, assignments are given which form part of the evaluation for the course. The type of assignments may vary. Students may be asked to write a report, conduct a small project, or conduct an experimental study. In a few cases, students are asked to assist in some research work. Site visits are sometimes organised, be it to look at the energy plant or to look at the effects of such plants on the environment. However, because of the constraints in finding a suitable time or site and enough funds, this is rarely carried out.

An exploratory study on the method of teaching at three universities was conducted. It was found that the most valuable part of an entire course depended on student involvement in such assignments.

Teaching Approaches via Project Work

Many courses now include the carrying out of a project. Individual students tackle a problem chosen by themselves or carry out work on research subjects assigned to them. Project work not only motivates the students, but also provides the opportunities for them to understand and apply the knowledge they learned from their lectures. This method is found to be effective, and a number of students become so encouraged that they continue their projects as part of their graduation exercise. The teaching strategy involves the following:

1. Allow students to chose a project related to energy.
2. Encourage students to talk to their lecturers about their proposed project in order to seek their guidance.
3. Divide the students into smaller groups. Each group is supervised by one lecturer or course tutor.
4. Students are allowed to describe or explain their projects to the group.
5. Encourage students to share what has been found with other students.
6. Each student must prepare a presentation of these findings, which include a research report, oral report, and a display or other visual presentation. In a project on design, students will be asked to describe each design in terms of its modifications and advantages over the traditional usage.
7. Students will be judged on the quality of the above reports.
8. Students are also given the option of writing a research paper on the design characteristics of alternative energy or the latest efforts to improve design. (The lecturers may assist in providing information about where to find important documents on this topic.)

As discussed above, project work of this nature not only motivates students to learn, but also allows students to use their knowledge. This activity gives the student an opportunity to choose a project that fits his or her own personal interests. By listening to other student presentations of their projects, students also gain an appreciation for the abilities and interests of their friends. Many students are surprised by the skills of those they have only known as their fellow lecture-mates. This method is effective when certain guidelines are followed. There should be proper planning, execution, and assessment in encouraging students to carry out project work. A proper environment, with tools or workshop facilities, should be provided.

Site Visits

A visit to a site provides a valuable experience for students. However, like going on any short expedition, it requires proper planning. Factors such as location, time involved, permission, transportation, and accommodations need to be considered well in advance.

Such visits can be arranged during the weekends or during the term-break. Sites suitable for visits are mining, processing, energy production plants, and research institutions. The focus should not be only on the plant but also the impacts of such activities on surrounding areas, e.g. on pollution. The visit provides first hand information and reinforces what they have learned with respect to social and technological relevance. Students can also gain some useful tips on vocational opportunities that are available. These sites provide openings for further investigations or projects.

Films

There are a number of films on energy which can be incorporated into the course. Such films as those on geothermal energy, deep-sea mining, radioactive wastes, nuclear energy plants, etc., are useful aids in teaching the related topics. However, prior preparation is necessary to consider when and for what purpose the films are used. Students should be guided as to what they are to look for while watching the films.

Inter-disciplinary Approaches

Co-operation is enhanced if lecturers from other disciplines help in the design and teaching of the courses and the courses are open to all students.

So far, in practice, the participation of lecturers from other disciplines in designing an inter-disciplinary course of this nature is yet to be seen. Each faculty has been willing to design such courses independently to suit its own orientation. The author found that there were so many similarities in the faculty courses that it would be more economical if they were integrated into one common core course.

18

Energy Studies

Y. B. KAMBLE

Community Science Centre, Ahmedabad, India

C. W. S. MURTHY

Institute of Technology, Chitradurga, India

J. M. FOWLER

National Science Teachers Association, Washington, U.S.A.

One Semester Course for Undergraduate Students

Prerequisite: Basic knowledge of mathematics and algebra at the high school level.

Objective: The energy picture over the world has changed a lot since the 1973 oil embargo, and all countries (both developed and underdeveloped) are aware of the limited resources and increased cost of oil and other natural energy resources. Research is being carried out in almost all countries to find alternate means for energy and its use. It is felt that all students need to know something about energy. Hence the main objective of these courses is to make science, non-science and engineering students aware of energy and its associated problems so that the students themselves can understand the issues involved in order to make personal decisions, and to be able to advise the local people in their area/field how to use energy in an efficient manner. Furthermore, the aim is to make the students aware that energy is essential for life and, by using it properly, human beings can lead comfortable lives.

The syllabi given below are general outlines and may vary from continent to continent and from country to country. It is hoped that the syllabi as outlined are useful to the instructors and that both the students and instructors derive benefit from them. While teaching, the instructors should emphasise the problems faced by their own countries and what steps are being taken to solve them. It is further suggested that the instructors provide illustrative examples and problems for students to work so that they will have some quantitative understanding of energy. Practical examples, demonstrations, and experiments should be incorporated whenever possible to carry students' interest in and understanding of energy beyond the course and thus enable them to help their countries in solving the energy crisis. It is best for the instructors to continuously evaluate student understanding. At the end of the term a fictitious country or region could be created and the students asked to study completely the energy problems in this region and suggest methods for meeting them. This paper could be treated as a term paper and made obligatory for all students to give to the instructor before taking final exams.

This course is intended for the final year of study.

1. Engineering students:

   The topics may be treated more rigorously and analytically. While teaching about limitation of energy sources, supply, etc., it is suggested that a lecture be arranged on aspects of energy management, for example, management of demand and supply. Some specific examples of a few industries/other systems may be given as exercises.

2. Science students:

   Although the topics to be treated are the same, they may be treated more from a conceptual point of view with a minimum of mathematics as students majoring in subjects other than physics or mathemetics are not expected to have strong mathematical backgrounds.

3. Non-science students:

   Though the topics are broadly the same, they need to be treated in a qualitative manner only, with more emphasis on environmental, social, and economical aspects. Simple examples of these aspects may be given. Care, however, needs to be taken to impart the basic concepts and information about the present energy situation and related issues.

Outline of Syllabus for Science and Engineering Students

A. *Energy—Introduction* (4 hours)

1. Historical and regional use of energy.
   - (a) Methods of energy generation—past and present.
   - (b) Usage of energy—past and present.
   - (c) Role of commercial and non-commercial energy.
2. Energy—manifestations and changes.
   - (a) Motion.
   - (b) Heat.
   - (c) Light.
   - (d) etc.
3. Forms of energy.
   - (a) Kinetic energy.

(b) Potential energy.
  (i) Gravitational.
  (ii) Chemical or electrical.
  (iii) Nuclear.
(c) Energy, force and work.
(d) Measurement of energy—SI units.

B. Conversion of Energy (6 hours)

1. Biological conversion.
   (a) Photosynthesis.
   (b) Food chains.
   (c) etc.

2. Other conversions.
   (a) Chemical to thermal.
   (b) Thermal to mechanical.
   (c) Mechanical to electrical.
   (d) Electrical to radiant.

3. Environmental effects and social impact of energy conversions.

4. Necessity for conserving environmental resources.

C. Laws of Energy and Efficiency (5 hours)

1. First law: Energy is neither created nor destroyed.

2. The efficiency of conversion.
   (a) Definition of efficiency.
   (b) Reasons for impossibility of 100 per cent efficiency.

3. Second law of thermodynamics—entropy.

D. Sources of Energy (12 hours)

1. Sources from the earth (descriptive only).
   (a) Coal.
   (b) Oil.
   (c) Natural gas.
   (d) Environmental effects of extraction and transportation.

2. Indirect solar energy.
   (a) Wood and biomass.
   (b) Hydropower.
   (c) Wind.
   (d) Tidal (non solar).

3. Direct solar.
   (a) Heating and cooling.
   (b) Electricity from photovoltaic and thermoelectric effects.
   (c) Solar ponds, etc.
   (d) Economic and manufacturing problems.

4. Nuclear.
   (a) Fission.
      (i) Technical description.
      (ii) Advantages and disadvantages.
   (b) Fusion.
      (i) Technical description.
      (ii) Advantages and disadvantages.

5. Environmental aspects.
   (a) Environmental effects.
   (b) Social impacts.

6. Focus on regional sources.
   (a) Consumption patterns.
   (b) Environmental changes.

E. ***Energy Utilisation*** (2 hours)

1. Commercial and non-commercial energy usage.
   (a) Buildings.
   (b) Transportation.
   (c) Industry.

2. Patterns of energy usage.
   (a) Graphs of regional and national consumption.
   (b) Illustrative examples.

H. ***Electrical Energy*** (8 hours)

1. Generation—AC and DC.

2. Transmission and distribution.

3. Utilisation of electricity.

4. Environmental aspects of generating plants.

5. Social impact of generating plants.

G. *Energy Limitations* (10 hours)

1. Resources.
   (a) Graphical representation of regional and world-wide usage.
   (b) Environmental and ecological effects.
   (c) Social impact.

2. Economics.
   (a) Real costs.
   (b) Hidden costs.
   (c) Regional and global examples.
   (d) Other implications like GNP, inflation, etc.

H. *Energy Conservation* (5 hours)

1. Conservation rationale.

2. Methods of conservation.
   (a) Reducing consumption.
   (b) Increasing device efficiency.
   (c) Increasing system efficiency.

3. Conservation strategies.
   (a) Regional examples.
   (b) International examples.

I. *New Sources and Technology* (2 hours)

1. Synthetic fuels.

2. Geothermal and other emerging technologies.
   (a) region-wide.
   (b) world-wide.

3. Other.

J. *The World Picture* (2 hours)

1. Pattern of consumption.

2. Resource distribution.

3. Technological adaptations.

4. Environmental impacts.

5. Social and ethical considerations.

Outline of Syllabus for Non-Science Students

Lecture 1: Energy and Its Importance (1 hour)

A. Topics:

1. Energy for survival and growth.
2. Energy the agent for change—biological, physical, or chemical.
3. How energy manifests itself—heat, light, etc.

B. Exercises/Projects:

1. List various types of energy utilised by an individual on a particular day, such as for cooking, transportation, etc.
2. Explain observed changes in terms of the forms of energy involved.

Lecture 2: Forms of Energy (1 hour)

A. Topics:

1. How different forms of energy are used.
2. Kinetic energy, heat, light, electricity, motion.
3. Potential energy or stored energy:
   (a) gravitational.
   (b) chemical/electrical.
   (c) nuclear, along with examples.

B. Excercises/Projects:

1. List various uses of electricity.
2. Identify common forms of energy used in the region, and list their rates of consumption in descending order.

Lecture 3: Conversion of Energy (2 hours)

A. Topics:

1. Conversion of energy from one form to another during physical, chemical or biological changes.
   (a) photosynthesis.

(b) food chains/degradation of energy.
(c) other conversions: electrical to mechanical, chemical to thermal, etc.

B. Exercises:

1. Name energy types that can be easily converted from one form to another. Which is most convenient?

2. Discuss how energy degrades during an energy chain.

Lecture 4: Laws of Energy (1 hour)

A. Topics:

1. Energy can neither be created nor destroyed.
   (a) evidence of 1st law.
   (b) efficiency of conversion.
   (c) work and energy.

B. Exercises:

1. Evaluate the efficiency of different energy systems.

2. Give examples of conservation of energy.

3. Provide examples showing that the amount of energy available after conversion is always less than before.

4. Demonstrate the 1st law: collisions, burning wood, driving a nail on wood with hammer, etc.

Lecture 5: Entropy and 2nd Law (2 hours)

A. Topics:

1. Entropy and 2nd law.
   (a) impossibility of 100 per cent conversion.
   (b) entropy as a general concept applicable to all systems.
   (c) order and disorder.

B. Exercises:

1. Show how entropy always increases in a system.

2. List examples of entropy increases.

Lecture 6: Measurement of Energy (1 hour)

A. Topics:

1. Entropy and 2nd law.
   (a) units of energy.
   (b) practical units of energy.
   (c) measurement of energy.

2. Cost of energy:
   (a) one unit of electricity in your area.
   (b) a litre of petrol, etc.

B. Exercises:

1. Compare how long different appliances will operate (e.g. a pocket calculator, a transistor radio, a toaster, a mixer, a refrigerator, etc.) with an input of one unit of energy.

2. Find calorific values of the food an individual consumed in a day.

3. Discuss changes in the cost of energy (food, fuel, etc.) over a decade; list.

Lecture 7: Consumption Patterns, Commercial and Non-commercial (2 hours)

A. Topics:

1. Energy in households and buildings.

2. Industries.

3. Transportation.

4. Agriculture.

5. Food.

6. Energy imbalance: national and world perspective.

B. Exercises:

1. List forms of energy commonly utilised in a particular region.

2. Diagram the consumption pattern in your household.

3. Identify the industries in your area and the fuel they use.

4. Compare the energy consumption pattern in your country with the world consumption pattern.

Lecture 8: Where does energy come from? Conventional Sources
(1 hour)

A. Topics:

1. Wood.
2. Cow dung/animal energy.
3. Coal/peat.
   (a) national resources of coal.
   (b) coal products.
4. Environmental aspects.

B. Exercises:

1. On the map of your country, show where coal is found.
2. Find:
   (a) How much coal is used per year in your country?
   (b) How much coal is imported?
   (c) How much coal is exported?
   (d) How is the environment affected by industries using coal?

Lecture 9: Conventional Sources 2—Oil (2 hours)

A. Topics:

1. Petroleum.
   (a) petroleum refining.
   (b) petroleum products.
2. Natural gas.
3. National resources and reserves.
4. Environmental aspects.

B. Exercises:

1. Show on the map of your country where petroleum or natural gas is found.
2. Determine national reserves.
3. Find the percentage increase in energy consumption over last year. At this rate per cent increase, calculate how long the national reserves will last.
4. Find how much oil is used for fuel. How much is used for other applications? How much is imported?
5. Visit a refinery.

Lecture 10: Nuclear Energy (2 hours)

A. Topics:

1. Nuclear energy concepts.
   (a) fission.
   (b) conventional reactors.
   (c) breeders.
2. Fusion and fusion reactors.
3. Environmental and economic aspects.
4. Radioisotopes for health and agriculture.
5. Limitations.

B. Exercises:

1. Answer the following:
   (a) Does your country have nuclear reactors? How many? What type?
   (b) List arguments for and against nuclear energy.
   (c) If your country has nuclear deposits, list what they are, and which deposits are used in reactors.
   (d) Discuss the use of nuclear energy in health and agriculture.

Lecture 11: Electrical Energy (2 hours)

A. Topics:

1. Electrical generation—general principles.
2. Steam and hydroelectric power stations.
3. Transmission and distribution.
4. Electricity in use: national picture.
5. Impact on environment.

B. Exercises:

1. Identify the use of electricity in various applications—agriculture, industries, transport, etc.
2. Determine the amount of power generated in your country.
3. Plot a graph of energy.
   (a) generation and
   (b) consumption in your country as a function of time for the data available from previous years.

Lecture 12: Renewable Energy Sources (2 hours)

A. Topics:

1. Need to look for non-conventional sources.

2. Commercial and non-commercial sources.

3. Indirect solar.
   (a) wood.
   (b) hydropower.
   (c) wind, etc.

4. Solar energy and its applications.

B. Exercises:

1. Determine traditionally non-commercial sources which have become commercial in your country.

2. Find the amount of
   (a) commercial and
   (b) non-commercial energy used in your household.

3. Name the renewable source which is most abundant in your country.

Lecture 13: Renewable Energy Sources 2 (1 hour)

A. Topics:

1. Biomass—food, fodder and fibre.
   (a) biogas.

2. Tidal.

3. Wave energy.

B. Exercises/Projects:

1. Visit a biogas plant, etc.

2. Perform simple experiments on biogas.

3. Determine how much cattle dung is required to produce 1 cubic metre of biogas.

4. List species of trees which should be grown in your region for fuel? What should be the characteristics of such trees?

Lecture 14: Limitations of Resources and Energy Use (1 hour)

A. Topics:

1. Resources.
2. Environmental impact.
   (a) extraction.
   (b) transportation.
   (c) conversion.

B. Exercises:

1. Describe the construction of a hydroplant and its implications.
2. Collect pollution data such as:
   (a) particulate matter.
   (b) oxides of sulphur, nitrogen, carbon, etc., and their effect on life.

Lecture 15: Limitations 2, Resources and Energy Use (2 hours)

A. Topics:

1. Social impact.
2. Economic impact.
   (a) real costs.
   (b) hidden costs.

B. Exercises:

1. Determine cost of extraction of coal/oil from deep layers.
2. Enumerate energy imports required at each stage in preparing toast/cookies. What cost is hidden?
3. Coal is transported from a coal field to a power plant to generate electricity. Calculate the cost of energy transport.

Lecture 16: Need to Conserve (2 hours)

A. Topics:

1. Ways to conserve.
   (a) cutting back.
   (b) increasing device efficiency.
   (c) increasing system efficiency.
2. Conservation strategies.
   (a) buildings.
   (b) transportation.
   (c) industry.

B. Exercises:

1. Calculate how much cooking gas/kerosene you will save per year in your household if you started using a solar cooker.
2. List some measures you could take to save energy at your:
   (a) house.
   (b) college.
3. Identify ways to save energy in transportation.
4. You are building a house. With the situation existing in your country, list various aspects you would consider to maximise energy savings.

Lecture 17: New Sources (1 hour)

A. Topics:

1. Synthetic fuels.
2. Geothermal.
3. MMD.
4. Hydrogen.

B. Exercises:

1. Discuss applicability of new resources to your country.
2. If your country has power generation plants using geothermal energy, visit and study the same. How much is used for heating purposes?

Lecture 18: The World Picture (1 hour)

A. Topics:

1. Patterns of consumption.
2. Resource distribution.
3. Technological adaptation.

B. Exercises:

1. Compare the national consumption pattern of your country with that of the world.
2. Discuss how life styles will be affected by the emergence of new technologies such as solar energy.
3. Illustrate graphically the *per capita* generation of commercial energy and consumption for different countries.

SECTION E

Energy Education around the World

Introduction

This section contains several reports written by the participants of the Bangalore Conference. These articles describe energy programmes in the participants' countries, or specific energy related participant activities, studies, teaching strategies and techniques, and other observations. Although site-specific in context, many of the objectives and activities can be adopted and/or adapted by others to another locale. Several of these articles contain mention of the emphasis placed upon energy education and the continued national development of energy programmes. Since the specific are often more interesting (actual case reports), this section begins with descriptions of specific programmes, and concludes with the more general 'what is going on in my country' type article. This was an editorial decision and was not meant to reflect which article or type of article is the more important. It was, however, based upon the premise that the intended audience for the volume would be teachers looking for specific ideas.

19

An Energy Module for Liberal Art Students

H. GOLDRING

Weizmann Institute, Israel

Science and technology are fast becoming an integral part of everyday life, and no one can afford to get along without understanding some of the basic principles in science, and their application in technology. As David Saxon[1]* puts it so aptly:

> ". . . The problem is that educated, intelligent, inquisitive people are unable consistently to bring informed judgment to bear on questions connected in almost any way to science and technology, questions often vital to the welfare of each of us and indeed to the future of the world. Instead the great majority of broadly educated people must rely on, and are at the mercy of, the testimony and assertion of others. . . ."

There is no doubt that science has to become an obligatory part of the high school curriculum, not only for science-track students but also for the so-called liberal arts students. This paper deals with a new physics module written especially for liberal arts students in the tenth or eleventh grade of the Israeli senior high school.

All students are required to take an integrated science course in junior high school (grades seven through nine, age 12 through 14), and to study physics at least one additional year in senior high school (grades nine through twelve, age 15 through 18).

The procedure adopted until recently has been to present the liberal arts students with a somewhat watered down version of the programmes taught to the science-track students.

* Superscript numbers refer to Notes at end of chapter.

Recently (1978), a committee appointed by the Ministry of Education, the members of which were physics teachers and physicists, recommended the preparation of special modules for liberal arts students. It was recommended that these modules be relevant to everyday life and be presented in an attractive way, without too many calculations, something in the vein of the "Science in Society" series.[2] Some of the modules the committee recommended be prepared are: Energy, Atoms and Nuclei, Acoustics and Music, the Universe, and Our Planet Earth.

The purpose of this paper is to describe the "Energy" module. In this module a survey of known energy sources is presented together with an outline of conventional ways of energy production and the problems encountered therein. Two novel ways of energy production are described. One is of general importance, from nuclear fuel by means of light water reactors. The other one, from solar energy, is of special importance to Israel, since this country has played a major part in pioneering new technologies in solar energy field.

This present module, entitled "Energy—the Problem, Sources, and Production",[3] is a direct sequel to "Electricity and Energy"[4] taught in grade nine of the junior high school in Israel. The module's outline described at the 1979 Rehovot meeting of GIREP,[5] and has since undergone quite a few changes which appeared necessary in the course of writing.

Some of these changes were made as a result of the analysis of a preliminary questionnaire administered to a sample population of about 200 students. Here we shall briefly mention just a few of them. A large percentage of the students were not aware that electricity is generated mainly by burning of coal or oil, and they thought that a nuclear reactor is a device for making bombs and other weapons. They had no idea how long fossil fuels are going to last at the present rate of consumption, and many were sure that nuclear fuel, i.e. uranium, will last forever. About half of the students did not distinguish between energy forms and energy sources, and they thought that there is a contradiction between the law of energy conservation and the energy crisis. This contradiction, however, did not seem to bother them. We tried to deal with all these misconceptions in this module.

The present version of the book contains six chapters:

1. Energy sources.
2. The energy problem: quantity and quality.
3. Electrical energy.
4. Solar energy.
5. Nuclear energy.
6. Energy—what price?

The first chapter follows closely the by now classical paper by M. King Hubbard: "Energy Resources in the Universe".[6] This chapter contains a brief survey of our renewable and non-renewable energy resources, the way they were formed, the estimated quantities in which they exist on earth, and the evergrowing rate at which they are used. Nuclear fuels are also briefly described. It is shown that almost all resources (except nuclear) can be traced back to the energy transmitted to us by solar radiation. But the energy radiated from the Sun is derived from nuclear reactions taking place inside the Sun. It thus transpires that all energy resources can be traced back to nuclear reactions (with the exception of tidal energy). Graphs describing the exponential growth of the consumption of the fossil fuels are shown, and the need for developing alternate sources is stressed.

The second chapter presents the concept of high and low quality energy sources. Heat engines are discussed, and the maximum theoretical efficiency of a heat engine is calculated. We do not go through the derivations of the Carnot cycle efficiency. We do try to make the result plausible, however, by arguments such as: Heat can only "flow" from one body to the other if there exists a temperature difference between them, and useful energy can be extracted from heat only when there is a flow of heat from a hot body (the hot reservoir) to a colder one (the cold reservoir). Furthermore, the higher the temperature difference between the two bodies, the more heat can flow from one to the other, and thus more useful energy can be extracted. We stress the fact that even with perfect insulation and frictionless movement of the parts of the engine, a certain percentage of the heat flowing between the two reservoirs cannot be converted into useful energy and must just flow to the cold reservoir, and thus is "wasted".

Since for practical reasons the highest temperature attainable in a steam turbine is about 600°C, and the lowest for the cold reservoir (ambient temperature) is about 20°C, it is easily shown that the energy percentage "wasted" as heat flow is at least 40 per cent, and usually more. It is stressed that this is not a shortcoming of technology, but an inherent feature of the conversion of heat to other forms of energy. In internal combustion engines higher temperatures are possible, but losses are greater due to the fact that linear motion is converted into circular motion.

As an example of a heat engine we introduce the conventional thermal power station. We calculate its theoretical maximum efficiency and compare it to the actual efficiency computed from data supplied by the Israeli electrical power company. In this way we hope to bring the rather abstract subject of thermal efficiency closer to the students. We then introduce the concept of the quality of an energy source and compare various known sources as to their qualities. We arrive at the conclusion that whereas the quantity of energy is always conserved, the quality of energy, however, is not conserved, but deteriorates. This chapter might appropriately be called: "the Second Law of Thermodynamics for pedestrians". In this chapter we show that the energy crisis is a result of "quality non-conservation".

The third chapter, Electrical Energy, was introduced because (a) this form of energy plays such an important role in our modern society and (b) the major percentage of electrical power is produced by heat engines (steam turbines). This is also a convenient way of introducing the concepts of a.c. voltage and current, and the transformer, which are lacking in the above mentioned ninth grade programme. In order to show the relevance of the subject to every-day experience, we supply data of the electricity consumption in Israel throughout the last ten years.

Having thus established that a shortage of conventional sources (i.e. fossil fuels) exists, and that a sizeable part of the energy problem is the production of electrical power, we then go on to deal with two as yet non-conventional electrical power sources: the Sun as a direct source, and nuclear fuel (uranium). As was pointed out earlier, these two subjects were chosen because nuclear energy is an issue of international importance and controversy, and solar energy is of particular importance to this country, being virtually the only energy source we do not have to import.

In the chapter on solar energy, the principle of operation of the saline solar pond is described in some detail. Israel has two solar ponds supplying electricity to the national grid. Both the 150 kW pilot plant at Ein Bokek and the 5 MW plant at Beth Haarava are located near the shore of the Dead Sea. This location is convenient because of the availability of salt water from the Dead Sea and because of the large percentage of sunny days per year in that area. The principle of the operation of the saline solar pond[7] is the creation of a temperature difference within the pond. A high temperature of 90°C at the bottom of the pond and a low temperature of from 20°C to 30°C at the top is caused by a salinity gradient (high salt concentration at the bottom and low at the top). Also described is an Israeli project to construct a pipeline through which water from the Mediterranean Sea will flow to the Dead Sea, 400 m below sea level. This project will enable the construction of a hydro-electric power plant as well as supply the amounts of water needed for topping up the solar ponds and for the cooling of conventional and nuclear power stations to be built in the future.

Another project introduced is that of the solar central receiver. "Solar One" in Barstow, California, is described as a working (10 MWe peak) example. Photovoltaic cells are also briefly mentioned.

Chapter 5 deals with nuclear energy. The principles of fission and fusion are introduced, and the Light Boiling Water Reactor operating with enriched uranium is described in some detail. Care is taken to stress the difference between an atom bomb and a nuclear reactor, because in the minds of students both are more or less similar. It is stressed that electricity production from nuclear fuel is the only method, at the present state of technology, which can supply the large amounts of power equivalent to those supplied by coal or oil powered stations and necessary for our subsistence.

The last chapter deals with the price one has to pay for energy, not only in money but also in exposure to technological hazards. There is no one hundred per cent safe method of power production, but of all existing methods we should choose the least hazardous. It turns out, by careful analysis,[8] that one of the least hazardous methods of power production is by means of nuclear reactors.

There is the need to diversify, however, and as Edward Teller[9] puts it so succinctly:

> ". . . No single prescription exists for a solution to the energy problem. Energy conservation is not enough. Petroleum is not enough. Coal is not enough. Nuclear energy is not enough. Solar energy and geothermal energy are not enough. New ideas and development will not be enough by themselves. Only the proper combination of all of these will suffice. . . ."

It is recommended that this module be taught in thirty periods; therefore, little time is available for students' experiments. We recommend that only demonstration experiments be carried out. Many good films and video tapes are available to accompany the subject. Field trips to a conventional power station, a solar pond, and a nuclear research reactor (Israel has no nuclear power station) are recommended. In conjunction with the module described here, and because of the special importance of solar power in this country, another unit dealing with experiments for utilising direct solar power has been developed by O. Kedem.[10]

The module is being taught this year in five classes, and we closely follow the progress of the students by classroom visitations and the administration of open and multiple choice questionnaires. We feel that the module fills an existing void in the curriculum, and hope that it will contribute in some measure to bringing our citizens up to date with a very important and relevant part of modern technology while teaching them a few principles of physics.

References

1. Saxon, David, The place of science and technology in the liberal arts curriculum, *Am. J. Phys.* **51** (1), Jan. 1983.
2. Lewis, John L., Project Director, Science in Society, Heinemann Educational Books, ASE, London, 1981.
3. Goldring, H., *Energy—the Problem, Sources and Production* (Hebrew), The Department of Science Teaching, Weizmann Institute of Science, 1983.
4. *Electricity and Energy* (Hebrew), Science Teaching Department, The Weizmann Institute of Science, Maalot, 1972.
5. Goldring, H., Energy—A 30 period module for mixed ability classes, Physics Teaching, *Proc. of the GIREP Conf.*, Rehovot, Israel, 1979, ed. U. Ganiel, Balban International Science Services, Jerusalem, 1980.
6. Hubbert, M. King, The energy resources of the Earth, *Sci. Am.* **224** (3), Sept. 1971.
7. Tabor, H., Solar ponds (Review Article), *Solar Energy* **27,** 181 (1981).
8. Beckmann, Peter, *The Health Hazards of NOT going Nuclear,* Golem Press, Boulder, CO, 1974.
9. Teller, Edward, *Energy from Heaven and Earth,* W. H. Freeman and Co., San Francisco, 1979.
10. Ganiel, U., and Kedem, O., Solar energy—How much do we receive?, *The Phys. Teach.* **21,** 573 (1983).

20

An Energy Programme for Sierra Leone Secondary Schools

E. LISK

Sierra Leone Grammar School, Freetown, Sierra Leone

A University of Sierra Leone/UNESCO National Training Workshop on Development of Physics Teaching Materials was conducted in Sierra Leone from 2 to 8 August, 1982. I was given the task of drawing up a programme for teaching about energy to pupils in forms 1 and 2 in Sierra Leone secondary schools. Pupils in these forms range in age from 10 to 12 years. This article contains a description of the programme which I developed and tested with members of my "energy team". The members of my team were secondary school teachers who taught General Science in forms 1, 2 and 3 and contributed greatly to the production and testing of the programme.

Introduction

This programme shows that energy is stored work. For mechanical energy, which is a common form, stored work is released when motion is halted or reduced. A body must be in motion before it can release the work it stores. Work is done by a body having kinetic energy, when its motion is halted or reduced. The teaching objectives, programme outline, and several related specific activities and relevant discussion material are given below.

Summary of Teaching Objectives

After going through this programme, pupils should be able to:

- relate work to energy and to define energy,
- distinguish the different forms of energy and their interconversions,

— recognise that mechanical energy appears in either kinetic or potential forms,
— identify these various forms of energy at home, in industry and in systems in which they occur.

Energy Program (outline)

Introduction

Energy—its meaning

Mechanical energy

Kinetic energy
— dependence upon both speed and mass

Potential energy
— dependence upon position in a force field
— gravitational and elastic

Internal energy

Internal energy and heat
Internal energy and mechanical energy
Internal energy and chemical energy

Energy conversions

Heat from mechanical energy
Electricity from chemical energy
Magnetic energy from electricity
Sound energy from mechanical energy

Evaluation

Activities

Kinetic Energy and Its Relation to Speed

Materials: Table tennis bat, retort stand with clamp, weights, a tennis ball.

Activity: A table tennis bat is held securely in a clamp attached to a retort stand. If no weight is placed on the base of the retort stand, then the stand will be easily knocked over by throwing a tennis ball onto the bat. A suitable weight is then placed on the base of the stand. For the retort stand to be knocked over, the ball needs to be thrown on the bat with greater speed.

Kinetic Energy and Its Relation to Mass

Materials: Two tennis balls, stones of different masses, pieces of string, a 3-m wooden pole, two flat rectangular boards, nails and a metric ruler.

Activity: A second experiment can be performed to show that kinetic energy is related to speed. A flat rectangular board is nailed to the end of a 3-m wooden pole. A second rectangular board of smaller size is nailed 20 cm below the postion of the first board. When the pole is upright, the larger board projects over the second board below it. A 3-cm nail is stuck on the front edge of each board. Each nail supports a pendulum made of a string and a tennis ball as the ball. The positions of the flat boards on the pole are such that when the pendulums are at rest, the tennis balls touch at their mid-sections; suitable lengths for the pendulums are 1.5 m. When one ball (B) is displaced from its rest position and then released, it will swing past its rest position. Ball B has been given a speed by allowing it to fall through a height measured above its rest position. The speed which B gains on reaching its rest position increases as it is raised higher and higher above the metric ruler, which is held horizontally just behind the tennis balls. Keeping the string that supports B taut, B is pulled to one side so that it is raised above its rest position. It is then released so that it falls on the other tennis ball (A). The experiment is repeated with B being raised higher and higher above its rest position. It is found that the displacement of A increases as B is raised higher and higher above its rest position. This simple experiment can show also, in a qualitative way, that kinetic energy depends upon mass. Six stones of different masses are used in turn as bob B and released, from a specific height, to strike tennis ball A. A moves farthest when struck by the stone of greatest mass. After these two experiments, the pupils may be able to see that kinetic energy depends on both mass and speed. And hence will accept that (verified by measurement) kinetic energy is half the product of mass and the square of the velocity.

Discussion

The type of energy so far considered is mechanical energy as opposed to heat or electrical energy. Mechanical energy confers upon a body the ability to move. In the case of kinetic energy, the body that possesses it is in motion. Whereas in the case of potential energy, there is temporary resistance to its motion which can be moved. Since a force is necessary to produce motion, a body will have energy if it is in a force field. A force field is a region inside which the influence of a force can be felt. An example of a force field is the earth's gravitational field. The earth's gravitational field is conservative. This means that the field causes a body to move in one direction but restricts its motion in the opposite direction to the same extent. This means that the work done, when a body moves in a closed path inside the field, is zero. The potential energy of a body in the earth's gravitational field (near the earth's surface) equals the product of its mass, its height above sea level, and the gravitational acceleration.

Elastic Potential Energy

Materials: A catapult made out of a forked branch, large rubber band, leather "holster", wooden block, and two G clamps; an unripened lemon fruit covered with an absorbent material; water.

Activity: Mechanical energy can be stored in a catapult as potential energy. The catapult is constructed out of the forked branch of a tree with the rubber tied to the ends of the forked stick. A holster made of leather is attached to the rubber band at its mid-portion. The catapult is mounted on a wooden block which is placed over a metre rule, and clamped firmly with two G clamps on a horizontal table. The missile is enclosed in the leather holster, and the rubber band is stretched beyond its natural length. When the tension in the rubber band is suddenly released, the missile is projected over a certain distance. The missile to be used is a small unripened lemon fruit which is not hard enough to hurt a pupil who may be struck by it. The small lemon is covered with a cotton cloth, the ends of which are then sewn together. The cotton material is wetted with water before the lemon is projected from the catapult. The cotton material will wet the spot where the lemon hits the ground so that the horizontal distance through which the lemon is projected can be measured. Each time the holster is pulled back, its position is noted on the metre ruler. This ensures that the extension of the rubber band is increased each time the experiment is repeated. The distance through which the lemon is projected increases with the extension of the rubber band. Because the catapult has done work on the lemon, the stretched rubber band has energy. Since the work done on the lemon is a measure of the potential energy stored in the rubber band, potential energy increases with extension. The potential energy in a stretched rubber band is half the product of the force stretching the band and the extension of the band measured in the direction of the force.

Internal Energy

Discussion

Closely allied to mechanical energy is internal energy. This type of energy does not confer upon the body as a whole the ability to move. Internal energy of a body is the total kinetic energy of the molecules together with their total potential energy resulting from the attractions and the repulsions they exert upon one another. The magnitude of the internal energy of a body can be increased by heating it or doing work on it. Since heat can increase the internal energy of a body, heat must be regarded as a form of energy.

Internal Energy and Mechanical Energy

Demonstration Experiment

If a metal bar is struck several times with a hammer, its temperature rises. This is simply because work has been done on the metal bar. The energy which this work has produced is internal energy. Since heat raises a body's internal energy, the temperature of a body rises when its internal energy increases. Another example that shows the relationship between heat and mechanical energy is rubbing two smooth stones together. The rubbed surfaces of both stones get hot. But what best shows the relationship between heat and mechanical energy is the seed of a certain fruit found in Sierra Leone. The seed is rubbed on the floor and applied to the hand; a burning sensation is felt on the hand. The skin can be injured in this way.

Internal Energy and Chemical Energy

Discussion

The internal energy of a compound determines its stability. Compounds with high internal energy are generally unstable and tend to dissociate, while compounds with low internal energy are generally stable. A chemical reaction will occur without the addition of heat energy if the total internal energy of the products is less than the total internal energy of the reactants. This loss in the internal energy usually results in the production of heat, but can be converted into electrical energy.

Electricity from Chemical Energy

Materials: Zinc rod, copper rod, beaker, dilute sulphuric acid, ammeter, 1.5-W electric bulb, connecting wires.

Activity: The simple galvanic cell shows the conversion of internal energy into electrical energy. In the galvanic cell, zinc and copper rods are dipping into dilute sulphuric acid inside a beaker. An ammeter is connected across the zinc and copper rods. The ammeter indicates a reading showing that electric current has been produced. The ammeter is replaced by a 1.5-W electric bulb. The lamp lights up and gets warm after some time. Here the internal energy released has been converted into electric energy, light, and heat energy. Thus all the four forms of energy are related. The 60-W bulb used in many homes lights up when electric current flows in its filament. These electric bulbs also get hot. Electric energy can also be converted into another form of energy called magnetic energy.

Conversion of Electrical to Magnetic Energy

Materials: Rectangular cardboard (30 cm × 15 cm), thick electrical cable, connecting wires with alligator clips, iron rod, wooden blocks to support cardboard, a thin piece of blade, a 3-volt battery.

Activity: A flat, rectangular cardboard with dimensions of 30 cm × 15 cm can be used to make a solenoid. Two parallel lines are drawn at the centre of the cardboard along its length, with a distance of 5 cm separating them. About twenty-five equally spaced holes are bored along each line, such that each hole on one line faces a hole on the other line. A thick electric cable is made to pass in turn through each of the fifth holes so as to form a solenoid. Both ends of the cable are made bare. A connecting wire is attached to each bare end of the cable by means of a alligator clip. The cardboard is fastened with drawing pins to two wooden blocks of the same size, so as to form a small table. An iron rod band and a half portion of a blade are placed inside the solenoid. When the iron rod is placed on the blade; it does not attract it. The connecting wires are connected to a 3-volt battery. The iron rod is again placed on the blade; it now picks up the blade. The experiment shows that electric current converts the iron rod into a magnet which therefore attracts the blade. The magnet must have energy if it can do work by lifting the blade. This type of energy is called magnetic energy.

Sound Energy from Mechanical Energy

Materials: Tuning fork, two nails at the ends of long piece of board, wire.

Activity: When a tuning fork is struck with a hard object, sound is produced. A close observation shows that the tuning fork vibrates as it emits sound. The sound ceases when the fork stops vibrating. This shows that the mechanical energy which the fork has as it vibrates is being converted into sound energy.

This conversion of mechanical energy into sound energy can be shown by another simple experiment. Two nails are struck at the end of a long piece of board. The ends of a thin wire are tied to the nails so that it is taut and above the board. When the wire is plucked sound is heard. The sound dies down when the wire ceases to vibrate.

Possible Suggestions for Evaluation

Here are some questions one can use:

1. What form of energy is basic to matter?
2. In the following systems, write P if the system possesses potential energy and K if it possesses kinetic energy.
   (a) A fully wound pendulum clock.
   (b) A rotating wheel.
   (c) A stretched rubber band.
   (d) A vertically projected bullet when it reaches its highest point.

3. Explain why a spark is usually produced when a very sharp instrument is used to chip a rock.
4. In the catapult experiment, does the kinetic energy imparted to the lemon equal the potential energy in the stretched elastic band?
5. State the energy changes which occur in an electric bulb when it is switched on.

General Comments

The materials used in the activities described above are cheap, and readily available. Most can easily be substituted by other things, for example, in the first activity a flat piece of board nailed to a wooden post stuck into the ground can serve as table tennis bat and retort stand. An unripened fruit such as lime or sour orange can be used instead of a tennis ball. A suitable substitute for tennis ball can be made from old newspaper. The newspapers are cut into small pieces and these are boiled in a solution of gum or glue. The semi-plastic thick mass can be rolled with both hands into a ball. The ball hardens into a solid mass as it dries up. This ball is light, but its weight can be increased by having a piece of stone embedded in the mass of paper. The ball also has bouncing property.

21

Site Visits as Part of Physics Education in the Senior Secondary School in Finland

K. HEISKANEN

Heinola, Finland

In 1982, all Finnish senior secondary schools incorporated a new curriculum which focuses on the interaction between school education and working life. The curriculum can be carried out at least in the teaching of physics and chemistry, and can be successfully incorporated into classes in history, social science and in student guidance.

During the class period pupils are introduced to the technological basis of industry, followed by an introduction to the working life, various professions, and the value, importance, and responsibilities of work itself. Pupils then become familiar with these subjects during on-site visits to industrial facilities, which have been carefully planned by the teacher. Following the visit students make group reports about the visit which are presented to the class and/or parents.

These site visits (field trips) can serve as an extremely effective way of bringing energy concepts into the curriculum. The chosen site could be a power production facility, or some energy intensive industrial site.

When planning to make a site visit, teachers might consider the following model which has proven to be quite successful in terms of an effective learning experience for students. This model contains the elements of teacher and student planning, critical observation, student report preparation and communication, and evaluation.

A Model for a Site Visit

Planning by the teacher includes:

— Identifying possible industrial sites for field trips. These sites should be within the local area in order to maximise the students' time on site.

— Selecting themes and subjects from schoolbooks and curriculum.
— Selecting the site most appropriate for theme and subject enhancement.
— Contacting the official at the site to get approval for the visit and to schedule the date of the visit. This contact may be made by phone. Making an appointment for a personal visit with the official to discuss the importance of this visit is often beneficial.
— Planning the pupils' visit together with the factory staff to determine suitable themes and subject areas is of extreme importance. If the persons responsible for co-ordinating the presentation and tour of the facilities does not know in advance of the teacher's instructional goals, the visit can be a disappointing experience.
— In-class preparation for the visit includes explaining and outlining for the students what they should see and/or experience during their trip to the site.
— Dividing pupils into groups. Each group is to focus on a different subject area.
— Assigning topics to the groups within their subject area.
— Dividing tasks within the groups. This activity is one where student planning occurs. Teachers should let each group do its own task assignments. Teachers should carefully monitor the groups' progress, however, providing suggestions and guidance only as needed.

The Visit

The teacher normally would assume a secondary role in this activity. The primary reason for the visit is to have student interact with site personnel in order to gain insights and understanding of the interaction of academic and working life. The visit should be planned so as to have the following occur:

— Introduction to the facility by site personnel and a description of the facility with supplemental diagrams, audiovisual presentations, and hand-out materials (if possible).
— Guided tour through the facility with adequate time for students to ask questions. Often times the industrial staff have difficulty understanding and responding to students' questions. The teacher should recognize this can (and probably will) happen and should be prepared to "interpret" questions and responses so that both staff and students can understand.
— Groups start to work on their assigned area as soon as possible. An ideal situation would be to allow them time on-site to discuss with their group what actions should be taken to meet their assignment goals. This is beneficial in that the students still have access to the industrial staff at this time.

Report Preparation

— Groups work to prepare their written and oral reports.
— Sometimes the groups may need to contact the staff of the facility for additional information.

Reports

— A group spokesperson presents the report to other groups, classes, parents, etc. There should be ample time for discussion after the reporting. The teacher should serve as the moderator for this discussion, with the group spokesperson as the responder with the opportunity of calling upon other members of the group for support and/or comment.
— The groups should be encouraged to make as professional a presentation as possible under time and financial constraints.
— Some reports may be worthy of disseminating outside the school environment. Care should be taken whenever this is done. Before distributing their reports outside the school, the industrial staff should be consulted to ensure report accuracy.

Evaluation

As in all educational activities, the teacher should obtain feedback on the students' experiences. The teacher should solicit responses from the students to questions such as:

— Was the visit a worthwhile activity?
— What could your teacher or facility staff have done to make the visit better?
— What could you have done to make the visit a better learning experience?
— [and other questions which probe student understanding]

Post-visit

Teachers should always express appreciation to all involved in making the visit possible. Teachers should also consider sharing any constructive suggestions for improvement of future visits at the site with site personnel.

Summary

In outlining the model above, several suggestions were made to help with planning, conducting, and evaluating a site visit to an industrial facility. These suggestions are not intended to be considered as all inclusive or encompassing in scope. They should be taken in light of the local situation and discarded or improved upon in order to make the site visit most effective in meeting the educational goals of the teachers and students.

22

Energy Education in Pre-vocational Courses in Ireland

R. MALONE

Trinity College, Dublin, Republic of Ireland

The course described below has been developed by a small group of practicing teachers supported by the CDVEC Curriculum Development Unit. Technical advice was provided by a physicist from a college of technology.

The course was developed within the context of the Irish educational system as shown in Fig. 1. Until 1969, secondary education was organised on a binary system: academic education in secondary school, vocational education in vocational schools. Since then, vocational schools have also provided academic education. A new type of school, the community school, was developed in the early 1970s to provide comprehensive education. Most secondary schools and some vocational schools are single sex institution. All community schools are co-educational.

The curriculum in second-level schools is controlled nationally by the government Department of Education. Syllabi are prescribed in "Rules and Programme for Secondary Schools". Examinations are also set by the Department of Education at 15 (Intermediate Certificate) and a 18 (Learning Certificate). The Learning Certificate also functions as a matriculation examination.

Some students opt to take technical pre-vocational courses at senior cycle. These are one-year courses and are certified by individual schools in accordance with criteria laid down by the Department of Education.

Science Education in Ireland

While Ireland uses advanced technology, the economy is largely based on agriculture and until recently there has been comparatively little science-based industry. Thus science is not seen as having a direct vocational relevance for the majority of students. Science syllabi are academic and geared mainly towards the needs of those who are likely to specialise in science or technology. Thus science is more likely to be considered an option rather than a core subject. "Schooling and Sex-Roles" shows that there are significant differences in provision and subject take-up between girls and boys, although science is taken by the majority of students, at least at junior cycle. "Science for All" has just begun to develop as an idea although the least able students may not be provided with science courses.

The Work of the CDVEC Curriculum Development Unit

This unit was set up in 1972. It is based at Trinity College, Dublin, and jointly managed by the City of Dublin Vocational Education Committee, Trinity College and the government Department of Education. The unit supports curriculum development and innovation in some sixty schools, mostly in the Dublin area. The work of the unit includes administration of curriculum projects in humanities, science, and outdoor education; alternative junior cycle courses for students opting for a vocational preparation programme. The unit co-ordinates the work of European Environmental Education network. A comprehensive account of curriculum development in Ireland since 1970 is given in *The Challenge of Change*.

Vocational Preparation and Training Programmes

As has been mentioned above, these programmes are provided as an option for students who have completed junior cycle of secondary education. They are one year programmes with three major components.

1. Technical studies (2/3). These may include one or more: engineering, electronic, construction, horticulture, catering or secretarial studies.
2. Personal and social development (1/3). These may include communications, social matters, science and society, industrial studies, and environmental studies. (The science and society models is at present provided only by a minority of schools. The energy module has been developed for this course.)
3. Students are also released from school for work experience.

Science and Society—Energy

The working group developed the course to meet the needs of a well-defined group of students; it had to meet the following criteria:

— It must be suitable for a range of abilities. A percentage of students following these courses have learning difficulties; most of the students are of average ability.

— It must be practically based.
— It must relate to the technical materials which the students are taking.
— It must be relevant to the everyday lives of students.
— It must further the personal development of students.
— It must involve the students in decision-making activities and thus prepare them for their role as citizens in a democracy.

The group decided to prepare a booklet for teachers giving sample student activities which could be based around existing published and readily available resources. The booklet is not designed as a student text which is followed from cover to cover. Instead, the teacher picks and chooses material suitable to the needs and abilities of individual students and refers them to published materials.

As most of the students in these courses will have followed a general science course at junior cycle, teachers wished to build on existing scientific knowledge and concepts. The booklet begins with revision of terms and definitions connected with energy. Energy was chosen as a topic because of its importance as a key unifying concept in all branches of science and technology and because of the obvious and well-documented social implications of the use and supply of energy resources.

In order to relate the concepts to the experience of students in the most direct way possible, activites were designed to begin with energy on a personal level, moving on to food–energy conversion. Consideration of energy use at home, in the community, and on national and global levels followed. The emphasis throughout was on using knowledge, concepts and skills to develop awareness of the importance of energy and of the implication of the use and limitations of energy resources, with consequent student willingness to modify behaviour in a positive way.

Methodology

The group of teachers who developed this course were all involved in an integrated science pilot project and were very committed to an activity-based problem solving approach.

They felt, however, that while this approach was necessary, it was not sufficient to meet the criteria listed above. They thus planned to use the media (newspapers, television, etc.) as a source of material for teaching and to use techniques such as group discussion, role playing, and decision-making. They also wished to integrate the course with the technical aspect of the students' programme in order to give the students a vehicle for the translation of scientific ideas into reality.

Science teachers would thus have to broaden their range of skills and become involved in areas traditionally associated with humanities teachers and technology teachers. Science also came to occupy a pivotal role in the curriculum (although occupying only a very small amount of a students' time). Teachers saw its potential as an integrator of the human and technical aspects of the student programme.

Future Plans

The booklet has now been tested in schools, and errors of fact and emphasis identified. Several construction projects have been prepared. Educators plan to develop and refine the material and, perhaps more importantly, to organise workshops for teachers so that they will able to realise the educational potential of this development.

References

1. Crooks, J. S. and McKebnan, J., *The Challenge of Change,* Institute of Public Administration, Dublin, 1984.
2. Hanna, D., *et al., Schooling and Sex-Roles: Sex Differences in Subject Provision and Student Choice in Irish Post-primary Schools,* The Economic and Social Research Institute, Dublin, 1983.
3. NFER, *Education Systems in EEC Countries.*
4. *Rules and Programme for Secondary Schools,* The Stationery Office, Dublin.

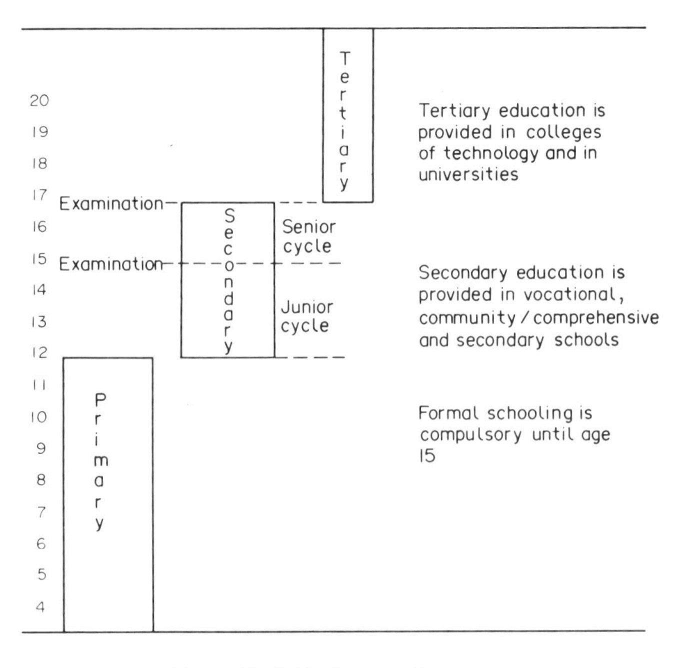

FIG. 1. The Irish educational system.

23

KSSP's Non-formal Educational Programme to Popularise the Energy Theme and Its Social Implications

M. P. GOVINDARAJAN

Chalavara High School, Kerala, India

T. P. SUKUMARAN

Kerala, India

Kerala Sastra Sahithia Parishad (KSSP) is a people's science movement in Kerala, India, dedicated to the popularisation of science and scientific thinking among the common people. The various issues pertaining to the ordinary man and his daily life, like problems of pollution, deforestation, and the unscientific aspects in the fields of education, health, energy, planning, and development have been taken up by KSSP.

KSSP entered the arena of popularising the "energy" theme in 1978 by bringing out a special issue of our *Sastragathi* science magazine. The various aspects of energy resources and the present energy crisis have been highlighted in it by eminent writers in the local vernacular. In course of time KSSP also brought out various pamphlets and popular science books like *Resources of Kerala*, highlighting the energy topic and its social implications. In 1983 another special issue of *Sastragathi* science magazine on energy was published,, which highlighted the problems of nuclear energy sources and other social implications of the energy policy of the government.

The topic "energy" has been made a subject of discussion among the masses in connection with the Silent Valley issue. The Silent Valley hydro project was introduced as a panacea for the energy crisis of Malabar, the northern part of Kerala. The installed capacity of the project was only 120 MW which supplies only a very meagre part of the energy requirement of the region. If implemented the project will eventually destroy about 30,000 hectares of valuable evergreen virgin forest which we cannot afford to lose. Hence KSSP launched a strong resistance to implementing the project and suggested various alternatives like the quick completion of hydro projects already started and the installation of thermal stations to meet the future needs.

The Improved Domestic Chulam Programme was another venture taken up by the KSSP to face the energy challenges. The major domestic consumption of energy in our country is in the form of firewood. Its yearly consumption for domestic purposes is estimated to be about 13,000,000 tons. This is one of the factors in the quick depletion of our forests. The cost of firewood is also increasing day by day. On realising this the research wing of KSSP with the co-operation of the committee on Science, Technology, and Environment of the government of India developed an improved domestic chulah which has about 33 per cent efficiency. KSSP is now popularising this high efficiency chulah among the people. This will help to reduce the firewood consumption of our state to a considerable extent.

Every year KSSP conducts Vinjan Pariksha (scholarship tests) for school children with the co-operation (which is not financial) of the Government of Kerala. Nearly 400,000 students of our state participate in these tests and we give scholarships to 300 students. This year "energy" is prescribed as the main subject for the test. In connection with this we have brought out a popular science book on energy for children.

The availability and proper utilisation of energy is a major factor in the development of any country. Any unscientific planning will jeopardise progress. This is what prompted us to take the issues to the people through informal education methods. For the same reason Energy should be made part of the formal school curriculum.

24

Teaching about Energy in the U.K.: A Physicist's View of the Changes

T. D. R. HICKSON

King's School, Worcester, U.K.

In 1972 the Nuffield Advanced Physics course became available in the U.K. This was designed for students aged 16+ and 17+, in their final two years at school. The course was highly innovative and it contained a unit on statistical thermodynamics entitled "Change and Chance". In it there was one section on "The Fuel Resources of the Earth" for which the time allowed was about 90 minutes. The impact made by this work both on the students *and* on their teachers was far out of proportion to the time it was given; it began a revolution in energy teaching.

Until that time, "energy" was learning to handle equations, learning about cunning experiments to show that "heat" and "energy" were equivalent and, occasionally, struggling to grasp some classical thermodynamics (*nobody* understood that!). The point is that all of this tended to be inward looking, to do with things in the laboratory. The real world outside was usually ignored. In any case, few of us who taught in schools had much experience of that world.

Then we began to consider the practical applications and social implications of an academic topic. We were forced to appreciate that real problems were not merely technical but also economic, and that sometimes the social and ethical issues may be even more important.

Physics teachers were, in any case, beginning to feel uneasy about what they were teaching. Although the "Learning through Discovery" courses, developed in the 1960s, had introduced an enormous amount of experimental work into the schools, they contained, for example, the dynamics of trolleys on runways rather than that of cars on real roads. We felt increasingly a need to be able to mention *applications* of what we were teaching. It was this, coupled with industry's increasing call for engineers, that led to the development of the Association for Science Education's "Science in Society" course.

This course, so much the work of John L. Lewis, had several aims but, in my view, the most important was to get into schools, and so into the hands of classroom teachers, new ideas and teaching strategies not readily available elsewhere. In this it seems to have succeeded, for copies of the material are to be found in nearly every school in the U.K.

The interactions among science, technology, economics, and society were studied under a number of headings. These were Health and Medicine, Population, Food and Agriculture, Mineral Resources, Energy, Facts (philosophy in disguise!), Industry in the Economy, Resources of Land and Water, and Looking to the Future.

To provide written material, at the right level, a large number of readers were produced. Students were required to critically evaluate written material, to debate, to evaluate consumption in their homes, to make visits (for example to quarries, industries, and coal mines), and to take part in decision-making exercises. These decision-making games were particularly successful. Supporting the "Energy" topic were three games (the Central Heating Project, Power Station Project, and Alternative Energy Project), which were simplified versions of exercises developed by Robert Gordon's Institute of Technology in Aberdeen for the Institution of Electrical Engineers.

As an example, the Power Station Project, like all of the decision-making games, is designed to occupy two or three forty minute periods. (Thus it can be slipped into a normal physics course; this has been done with pupils at the 14+, 16+ and 17+ stages.)

The project is a simulated case study dealing with the various decisions that have to be made when a new power station is being planned. It is assumed that an Electricity Generating Board is to build a new 2000-MW power station in a certain (imaginary) area and that the object is to decide which type of station should be built—coal, oil, or nuclear—and where it should be sited.

First, pupils are prepared either by a preliminary lesson on the basic principles of power generation or by reading a paper specially written to cover this material. Then they are divided into three groups which are each allocated their particular type of power station. Each group, working co-operatively, evaluate six possible sites and then decide on the best for their type of power station. At the next stage, each group splits into two sub-groups, one to produce a detailed map of the site for their scheme, the other to calculate the various costs involved.

Finally, the three groups, in turn, report their findings to the class as a whole so that a general discussion can follow as to the relative merits and disadvantages of each scheme. Then a decision has to be made. This is best done by bringing in another adult, preferably with some experience of making decisions on complex matters, as an assessor.

It has been my experience, that students find this type of exercise highly stimulating and their increased awareness of the need to make reasoned decisions that take account of all relevant constraints affects their attitudes toward the complex problems they may meet thereafter. Once schools have had experience with these decision-making games, they may well be motivated to develop some of their own.

In all of this work, teachers are faced with the problem of where to start: should they begin with world, national or local issues? In spite of the growing awareness of others shown by today's students, our experience has generally been that it is better to start with what they are familiar—their homes—and then to work outwards. This gives them more confidence and helps to prevent them from offering opinions which are unsupported by facts. They need guidance if they are to avoid making proposals based only on prejudice or misinformation.

In recent years schools in the U.K. have had available to them a growing amount of information on energy matters. Indeed there has been such a flood that teachers have found it difficult to keep up to date and their pupils interest has tended to become saturated. (In 1972, students sat in shocked silence as the finite nature of our planet was revealed to them. Now as yet more "gloom and doom" emerges from television and radio, they just yawn.)

There is much free material from oil companies and from the U.K. Atomic Energy Authority and, at the other end of the spectrum, from pressure groups such as Friends of the Earth and Green Peace. It is vital that students should be required to compare and to contrast such material critically. Television programmes are often visually stimulating although the accompanying discussions are all too often superficial. However, there are films, such as British Petroleum's *Energy in Perspective,* and videos, such as New Scientist's *Energy,* which are very useful to use to generate discussions. Becoming available are computer simulations of nuclear reactors, for example, and decision-making exercises based on computer programmes, particularly in the field of energy conservation. Computer data banks are also beginning to be used in schools.

However, when there is a major review of the most recent information, such as occurs at the World Energy Conference, this is not available in schools. It is to the credit of the organisers of this conference that we are setting out to put into schools the ideas collected in Bangalore.

It is a sign of the times that recent physics courses designed for the age range 11–16 contain sections on Energy and Power, on World Uses of Energy, and on Nuclear Energy. This year a revised version has been published of the Nuffield Advance Physics course mentioned at the start of this paper. As well as a revised unit on statistical thermodynamics, entitled "Energy and Entropy", there is a whole new unit on "Energy Sources". This new unit deals with The Background to Energy Supply and Demand, Power from Nuclear Fission, Conserving Fuel in the Home, and Energy Options. Also published, to support this chapter, are two background books on energy.

Increasingly, the public examinations, taken at the ages of 16 and 18, contain questions on energy topics. Topics covered include the generation of electricity from wind and solar energy, energy efficiencies of petrol and battery powered cars, and the Laws of Thermodynamics applied to power stations.

One of the most useful teaching strategies I have come across in recent years depends on a tape–slide programme. It is produced by WWF/IUCN International Education Project and is a WWF India Presentation called *Vanishing Forests*. It deals with the problems of deforestation and some proposed solutions. If this programme had been produced for U.K. audiences it would have had a British voice giving the commentary. Because the voice is Indian and because the material was made for Indians to show other Indians (for U.K. students it is like being "a fly on the wall"), the problem is obviously real and its solution is clearly urgent. Students find the programme quite fascinating. It also helps to make what seems to me to be a most important point, that although our students must be shown the problems which we face, they must not view the future with pessimism, but rather as a challenge.

For many years now, the wide estuary of the River Severn has been considered as the site for a tidal barrage. Recently, there have been detailed reports on the feasibility of such a scheme made to the British Parliament. As with all such reports made for our politicians, the language has been kept simple and so our students can read and understand them too. At my school we held a successful seminar involving the Departments of Physics, Biology, Geography and Economics. Students were briefed to speak about the barrage from the point of view of their own subjects and then the group as a whole discussed the scheme.

The topic for the next term's seminar is "Energy in Agriculture". Here we hope the chemists will join us. These exercises encourage students to realise that their future contribution to the solution of complex problems will probably be as members of a team. They seem less likely then to suffer from the apathy born of frustration. ("What can I do, on my own?")

We have all become used to the concept of the Energy Crisis; indeed the problem is often to know *which* crisis. By studying energy, students can be led to see that the future is full of exciting challenges for those with a scientific or technical training. They can then, perhaps, understand why the Chinese symbol for "crisis" includes the symbol for "opportunity".

25

Energy Education in Norwegian Schools

H. HARNAES

University of Oslo, Norway

Briefly about the Norwegian school system

In Norway, children have to go to school for nine years. They start when they are 7 years old. Everybody has the same curriculum.

The upper secondary school is not obligatory but 85 per cent of Norwegian 16-year-olds continue in the school system: 40 per cent on theoretical tracks and 45 per cent on vocational tracks.

Norwegian Energy Resources

The energy situation in Norway is very unique:

— Excluding communication and transport, 100 per cent of our energy needs is supplied from developed waterfalls. Even energy-intensive industrial operations, such as aluminium production, get their energy from hydro-electric power stations.

— We export electric energy to our neighbouring countries.

— Norway has oil and natural gas in the North Sea. We export both oil and gas.

This unique situation results in discussions whether it is right, with respect to people's need for recreation and/or from an environmental point of view, to develop more waterfalls. Such discussions are of course meaningful and important, but the problems may be regarded as luxury problems for other countries.

Another consequence of our privileged energy situation is that we have not yet had to face a serious discussion about nuclear power plants.

When and Where Energy is Taught

On the primary level the pupils have an interdisciplinary subject which contains science, social science, geography, and history. Pupil activity and social aspects are stressed in Norwegian school advisory plans. But in fact the Norwegian teachers can decide the content themselves.

On the lower secondary level energy is taught both in science and in social science. The physical aspects of energy have traditionally been taught in the science lessons. The social aspects—energy crisis, energy resources, energy in the homes, etc.—have, on the other hand, been taught in social sciences. This is now changing.

The theoretical tracks on the upper secondary level have a common first year with science as one of the subjects. The pupils either specialise in physics, biology, or chemistry the second and third year, or they have no more science. The social aspects of energy are taught in the social sciences.

In the vocational tracks an effort is made to make science more relevant to students' trades, which differs from department to department and from school to school.

New Trends in Science Education

Science has traditionally been taught without influence from the social sciences. The new plans from the national board of education recommend more interdisciplinary education. While energy has been a favourite topic for different projects, the scientific aspects of energy have often been left out.

In the future we hope that science teachers will also deal with social aspects and that scientific aspects will be a natural part of interdisciplinary projects about energy.

Trends in Physics Education

There is a trend in Norwegian schools towards increased pupil activity in physics lessons and fewer demonstrations done by the teacher. There is therefore an increased need for simple, low-cost equipment.

The Norwegian project "Girls and Physics" has, together with similar projects in other countries, shown us that there is a correlation between sex roles and pupil achievement in physics. Boys' experiences with cars, trains, mechanical kits, batteries, etc., make them achieve better in physics, especially in mechanics and electricity. A consequence of these findings is that schools must now provide girls with these same experiences, so that girls can get rid of their "apparatus phobia". Both girls and boys will benefit from these activities, especially in countries where technology is not an ordinary part of the pupils' daily surroundings.

Trends in Energy Education

In the field of energy many suggestions for similar pupil activities have been developed. The Centre for Science Education, University of Oslo, has been leading this development.

Step one in energy education has been that pupils learn the concepts, energy source, energy receiver and energy chain, and that they learn to look for indications of energy transmission. Also, an important part of the teaching aim is that pupils learn to observe and reflect. Step two in energy education deals with different forms of energy, work, and power.

26

Energy Education in Venezuela

J. A. RODRIGUEZ

Caracas, Venezuela

The development of the concept of energy is one of the greatest conquests of the human mind. Not only does it greatly simplify the description of the material world; but also, the conservation of energy principle is one of the fundamental principles governing all phenomena known to date.

Teaching about energy in Venezuela is carried out in the traditional way. However, since 1978, there has been a tendency to give it more importance at all educational levels. This tendency originated in and is supported by the National Centre for the Improvement of the Teaching of Science (CENAMEC) through various projects. In 1978, CENAMEC brought Dr. Gideon Carmi from Israel to train a group of primary school teachers for whom some materials pertaining to the Petach Project were translated and adapted. In this training, much importance was given to the teaching of the energy concept in primary school.

Other projects are under way. The application of the project "Education in Energy in Secondary School", sponsored by CENAMEC and the Ministry of Energy and Mines, began in 1982. At the present time, efforts are being made in CENAMEC to start the project "Education in Science Technology and Future Human Needs". This project, to begin in 1986, would make curricular recommendations and test some materials by inserting them into the curriculum.

Education in Primary School

Teaching the energy concept can be started in kindergarten and continued in primary school, taking into account the stages and sequence of the children's psychological development.

It is important that children become familiar with the qualitative aspects of energy in kindergarten and the first year of primary school by playing with objects that have energy (sources of energy) and which pass it on to other objects that transform it (energy transformers). Rubber bands, springs, balloons, toy battery powered motors, etc., are all good examples of energy sources to be used at this level. For energy transformers, children can use pinwheels, and spring, rubber band, or battery powered toys such as cars, tractors, boats, and ejectors.

In the second year, they can make estimates and establish comparisons as preparatory training in rudimentary energy measuring. It is advisable to initiate children in the work concept by, for example, enlarging springs or moving bodies on a horizontal surface. It is important that the children realise that every time they work there is a transfer of energy.

In the third year, children can effect rudimentary energy measurements by using any of the following devices: a wooden bead inserted in thin wire and a ruler, a toboggan, and a wooden block and ruler.

The formalisation of the concept of energy may be initiated in the seventh year. It is necessary to be precise in identifying and conceptualising cause-effect relationships, the energy transformation chain, and generalisations.

Education in Energy in Secondary School

The first step would be to study different forms of energy and their application to everyday life. The second step would encompass the study of mechanical work as a means to transfer energy. The third step would include the study of the following subjects: kinetic energy, gravitational potential energy, and caloric energy; the equations for the different types of energy, their units and their equivalence; transformation of work into heat and vice versa; principles of thermodynamics; and entropy (order and disorder).

The subject "Energy in the House" should consider artifacts and utensils that create "consumption", the estimation of costs, and the solution of "consumption" problems. It should also be explained that what we call "consumption" is only "transformation". Topics, such as "Energy in Industry" and "Energy in Transport and Communications" should be discussed.

Since Venezuela is an oil-producing country, marketing and contamination problems regarding this product should be analysed.

Education in Energy at Tertiary Level

Special attention should be given to the subject "Energy and Society"—energy as a parameter of economical and social development. The subject "Energy and Industry" should be subsequently considered, with emphasis on the different energy resources used and their environmental effects. Ways to control contamination and still maintain the equilibrium of the renewable resources used should be discussed. Alternative forms of energy as substitutes for non-renewable resources should also be treated. The subject "Transportation and Communications" should be extensively examined.

Both social and economic development, and influencing factors, should be approached, giving attention to the subject "Independence and Energy". Since oil is Venezuela's main source of income, any expense implying foreign exchange affects the quantity of oil sold. Should we decide to take further steps to reduce consumption, this would mean a considerable saving of foreign exchange and an enormous saving of oil, which, in the end, would result in a healthy conservation policy. However, Venezuela relies on its oil and, for this reason, an adjustment is necessary.

27

Teaching Energy for Social Needs

T. P. SUKUMARAN

Kerala, India

M. P. GOVINDARAJAN

Kerala, India

Energy has often been a subject of discussion in development conferences and among scientists and planners in our country. However, no serious thought has ever been given to make it a discussion among educators. As such it has remained a boring and tiresome topic in the general science curriculum.

Life itself can be defined as exchange of energy between the living beings and the medium that surrounds them. Energy should be taught both as a concept and as part of daily life. Hence the curriculum itself should be framed to meet the challenges and needs of the future.

Topics

The new topics to be introduced in the curriculum should include the following:

1. Concepts: What is energy?: matter and energy, different forms of energy (mechanical, heat, electricity, light, sound, atomic, solar, etc.)

2. Sources of energy:

    A. Sun, as a primary source: food and muscular energy, firewood, coal, natural gases, petroleum, biogas and cow dung, wind energy, ocean wave energy, hydro energy.
    B. Sun, as a direct source: solar heaters, solar pumps, solar dryers, photovoltaic cells.
    C. Geothermal energy.
    D. Tidal energy.

E. Atom, as a source: mass—energy equivalence, atomic reactors, atomic energy (problems and promises), thermonuclear process, plasma, limitations.

3. Energy exchanges:

A. Living system: flow of energy through plants and animals, food web, human and animal labour, food to muscular to mechanical energy.

B. Non-living systems, which includes experiences from daily lives:

1. Domestic chulahs (wood burning stoves) —chemical to heat
2. Electric stoves —electrical to thermal
3. Loud speakers —electrical to acoustical
4. Electrical bulbs —electrical to light and thermal
5. Microphones —acoustical to electrical
6. Cells —chemical to electrical
7. Heat engines —thermal to mechanical
8. Hydro stations —mechanical to electrical
9. Thermo stations —thermal to electrical
10. Dynamo —magnetic to electrical
11. Motors —electrical to mechanical

4. Energy crisis—domestic, regional, national, and global:

A. Depletion of natural resources.

B. Factors responsible for depletion.

These four topics should form the core of the energy study in the secondary level. The teacher should be capable of introducing more examples from local surroundings.

Objectives

1. The students will understand the importance of energy in their daily lives. They will appreciate that their growth, both physical and intellectual, takes place purely by the exchange of energy between themselves and their suroundings.
2. The students will be aware of the importance of energy in living things, the relationship between living and non-living in the matter or energy, and the relationship of various non-living things in the matter of energy. It is energy that binds living and non-living things.
3. The students will correlate the learning experience obtained in school with the domestic environment.

4. The students will understand the present energy crisis at the domestic, regional, national and global levels.
5. The students will understand the importance of the conservation of energy sources with emphasis on the menace of overuse. The challenges and possibilities of the future will be pointed out.

Methods

1. School and home: List all the energy transformations taking place in the school. The students will then correlate the activities in the school with those taking place in the home so that the activities in the home will turn into a learning experience.
2. School and neighbouring energy transformation centres: The students will go on study visits to the neighbouring energy transformation centres like a hydro station, thermal station, mechanical work shops, etc. This should be made part of the teaching–learning activity.
3. The energy study at present can be augmented by national planning and programmes by developing twenty to thirty teaching programmes, each 15 minutes long, to be shown through the television systems. Our task will be to divide, formulate, and prepare these programmes. These may be telecast daily at certain regular intervals. After watching each programme the teacher will relate the programme's topics to the local situation. The teachers must be given suitable orientation courses so that they are capable of handling the new curriculum.

28

Teaching Strategy for the Effective Implementation of Energy Education

B. G. KUSUMA

Acharya Pathasala Girls High School, Bangalore, India

Energy is essential for existence. Hence it is necessary that energy education be imparted to children who are to be the citizens of tomorrow. To be effective, implementation should start in the primary classes and continue through the secondary and tertiary levels, apart from educating the general public. This can be done by adopting an ever-expanding curriculum in the schools.

The content part of the curriculum should include:

1. The concept of energy.
2. The different sources of energy.
3. The ability of energy to change form.
4. The resources now used in rural and urban India and also in the world.
5. The energy resources of the future.

The teaching strategy used for effective education depends on:

1. The group that is taught.
2. The size of the class.
3. The socio-economic status of the taught.
4. The particular topic.
5. The available instructional media.

The same is true in the case of energy education. Here, a scientific approach in presenting the topics will be more effective. The process of science in solving a problem includes, as all of us know:

1. Making observations.
2. Collecting information.
3. Proposing, testing, and evaluating hypotheses.

In India a marked difference can be noticed in the socio-economic and academic background of the urban and the rural pupils, though the curriculum taught is the same. Moreover, in the urban area the classes are generally heterogeneous. Such a state of affairs necessitates the adopting of different strategies for different topics. The following media/technique have been found useful and effective:

1. Discussion.
2. Providing reading materials.
3. Providing self-learning materials.
4. Films.
5. Visit to national science labs and museums.
6. Dramatisation.
7. Tape and slide lessons.

To teach the different sources of energy, self-learning materials and reading materials can be provided. Energy resources and their uses in rural and urban areas can be taught through charts and wall magazines. Nuclear energy is one of the main energy sources of the future. I have made an attempt to combine two of these media, that is, a film show and a discussion following it. The following is a brief description of the procedure followed.

Example 1

A group of students of the IX Standard class were taken to the Visweswariah Industrial and Technological Museum to see the film *Atomic Energy*. It is a 13-minute film produced by the Energy Encyclopaedia Britannica Educational Corporation, collaborator being Mr. Albert V. Baez.

Synopsis of the film

Though atoms are invisible, the effects they produce can be seen. White streaks similar to vapour trails left by jet planes are made in a cloud chamber by atomic particles, 10,000 times smaller than the atom itself. In this film, the structure of the atom is revealed through the use of animation. The atom's effects are shown through demonstrations involving a cloud chamber and a Geiger counter. Its uses are shown in hospitals, atomic power plants, and atomic submarines.

Objectives in showing the film

1. To illustrate the structure of an atom.
2. To illustrate the behaviour of radioactive atomic particles.
3. The uses of radioactive materials.

4. To help the pupils to develop the ability to make accurate observations.
5. To motivate the students to learn more about nuclear energy.

After the film showing, the students' reaction were observed. All were eager to talk about what they had seen and learned about the atom. Some of the questions given to the students are listed below:

1. What is the nucleus of an atom?
2. How was the nucleus shown in the film?
3. How does the nucleus of radioactive element differ from that of a non-radioactive element?
4. What are some uses of atomic energy?

Example 2

Research and discussion were used to explore the social significance of atomic energy. The discussion related research knowledge to what was already known. This enabled the students to clarify their understanding of nuclear energy.

Plan of the discussion

1. *Topic:* Nuclear energy: What is it? What are its uses and abuses?
2. *Duration:* Four class periods.
3. *Participants:* IX Standard students.
4. *Operation:* The class was divided into groups. Group leaders were selected. The pupils were given three class days to gather reference information. The group leaders spoke in the discussion while the others were active listeners. The discussion was tape recorded.

The class participated in this discussion with full enthusiasm. A sample of questions raised during the discussion are given below:

1. Do you know a natural source of nuclear energy?
2. How can man generate nuclear energy?
3. Where are the nuclear power plants in India?
4. What happens if the nuclear energy production in a reactor is not controlled?
5. What are the harms that a nuclear reactor may possibly cause?

Many more questions were raised and answered satisfactorily. The pupils were able to distinguish between fusion and fission. They seriously considered questions such as "Suppose a neighbouring nation prepares a nuclear bomb, what should we do?"

They were able to correctly think of another possible source of energy in our country. They answered the question "Can you mention any other energy resource which may be used in the future?", with responses such as "solar energy". This strategy for this particular topic did achieve the following objectives:

1. To instill an interest in learning.
2. To collect facts and information from different books and magazines.
3. To become aware of national and international problems relating to the energy possibilities.

As already mentioned, there can be more than one strategy to impart energy education effectively. Colleagues and other teachers in Karnataka are trying their best to use the different strategies.

India is a rurally based nation. For effective implementation of energy education, we cannot ignore our villagers. The energy resources in rural India are somewhat different from what they are in the urban areas. Therefore, energy education to villagers must be planned so that it would be possible to get the maximum benefit out of the local resources.

Animal and human labour (i.e. muscular power), firewood, crop residues and animal waste like cow dung form main resources of energy in rural India. Men and women of our villages toil too much and all the energy they get is by their own physical efforts. Progress of rural India could be brought about by increasing the quantity and quality of energy use.

To achieve this aim, it is imperative that effective energy education be imparted to the villagers, particularly to women, who cook the food, heat the water, do other household work, and also go out to work in the agricultural fields. Both types of work consume energy. Therefore, energy education (that is knowledge of better utilisation and greater conservation of energy) is a must for women. The efficiency of the fuels used for cooking and heating can be increased by the use of improved stoves. Installation of a community biogas plant provides both manure and fuel gas. Solar energy should be harnessed wherever the climate is suitable. Thus villages can be made self-sufficient in this respect.

"Astra" stoves and gobar gas plants are steps in this direction. But to increase their popularity, it is essential that villagers be educated. This education can be informal. The strategy is to make use of mass-media like documentary film shows, radio programmes and dramatisations.

"Energy" is a complex topic. In fact it is the axis around which rotates the wheel of life. Energy is interpreted by different people in different ways. For example, to a physicist, energy is the ability to work. A biologist, a politician, and the common man have their own ideas about energy. All these have to be taken together; they are complementary to one another. For effective energy education, there must be an inter-disciplinary teaching team and a correlated curriculum. An ideal energy education should reflect the salient aspects of our energy situation, its multi-disciplinary nature, a sound scientific approach, careful consideration of alternative sources and their impact on society and above all an urgent call to action.

SECTION F

Sample Energy Education Fact Sheets

Introduction

This section contains sample Energy Fact Sheets of the kind previously developed and distributed by the U.S. National Science Teachers Association under the project management of John Fowler. Dr. Fowler was the team leader for the tertiary group at the Bangalore Conference and undertook the task of preparing updated versions of some of the more appropriate energy source fact sheets. These are to serve as samples which energy educators can adapt to their own specific climatic, energy resource, and educational needs. Although each "fact sheet" gives the resource availability in the United States, the intent is for energy educators to replace the U.S. specific data by national or local data from their own countries. This adaptation would serve to make the material more relevant to student needs and therby increase student interest. Many of the "fact sheets" produced by NSTA are outdated in terms of resource consumption rates and availability. Some of them are still available and could be updated by interested individuals or agencies.

29

Solar Heating and Cooling

J. M. FOWLER

NSTA, Washington, U.S.A.

Humankind has always looked to the Sun for warmth. It was our only source of heat during those hundreds of thousands of years of fireless prehistory and, since fires themselves are so inefficient, was still the major source of heat until the real beginnings of the industrial age 200–300 years ago.

Primitive solar heating was relatively technology free. Our ancestors took the Sun's heat when and where they could get it. In the past few generations, however, our expectations of comfort have changed and we have become accustomed to a dwelling that is warm throughout, not just around the fireplace. In the warmer, affluent countries we also expect summertime air conditioning of our houses, offices, cars, and even apparently our athletic stadiums. We not only want all of the buildings heated or cooled, but also want immediate control over the temperature. Consequently, we must rely on electrical fuel-burning furnaces, air conditioners, and hot water heaters.

We still use solar heat around the world to, for example, dry clothes, extract salt from sea water, bake bricks, and dry food to preserve it.

Solar water and building heating and cooling was under development long before the Oil embargo. Some 30,000 or so solar hot water heaters were in use in the southern United States in the 1930s and 1940s. Solar evaporative coolers are used in many countries. But the rapid rise in the price of energy from fossil fuels created new interest in this energy from the Sun.

Technology

All solar technologies are constrained by several unique characteristics of solar energy. It is a large resource, but is spread thinly over the Earth's surface. As we point out later, it would take a square metre sized collector operating at 100 per cent efficiency to keep a 200-watt light bulb burning.

Solar energy is also erratic. It varies with latitude as the angle of the Sun's rays with the Earth's surface changes; it is strongest at the equator and weakest at the poles. It also varies regularly with day and night and irregularly with cloud cover.

The two major problems that the technology must solve are concentration and storage. We discuss technological responses to these constraints for solar space and water heating and solar cooling below.

Styles of Solar Space Heating

Styles of solar space heating can be categorised as either "active" or "passive". In an active solar house, energy is collected by a system of solar collectors, typically on the roof. The heat energy is carried away by water or air and transferred into some large storage system. From the storage unit, it is then circulated, as required, throughout the house by either a hot water or hot air system of conventional design. Active systems require additional energy input in order to operate the circulation and distribution system.

Passive Systems

In a passive system there is no circulation to storage. Passive solar houses are most useful in generally sunny regions. They tend to be custom designed, energy efficient in all ways. They could properly be called "Solar Conservation Systems".

Solar heating in passive solar homes is both direct and indirect. Double or triple glazed windows admit heat to a room or to a "greenhouse" added to the sunny side of the house. Heat is absorbed and stored in thick floors or walls which then give back the stored heat at night.

In indirect heating, the sunlight first strikes and is absorbed in a large thermal mass. This mass then re-radiates the heat later when it is needed. The Trombe wall is a well-known example of an indirect heating device. A thick dark masonry wall protected by double glazed windows faces the Sun. Cold air from the room to be heated flows through vents in the bottom of the wall, is heated as it passes up along the wall, and enters the room through top vents.

Active Systems

In an active system there are three basic components, the collector system, the storage system, and the distribution system. There is considerable variety in the devices and materials used in these three systems.

Collectors

The "flat plate" collector is the most common, a large, shallow box covered by a transparent plate of glass or plastic. The inside is blackened to absorb solar radiation, and the heat that is generated in this absorber is carried away by water (in pipes or trickling across the surface) or by the air. In sophisticated systems the interior is a partial vacuum to reduce energy losses, and the windows are double layered or specially coated to admit incoming radiation but trap outgoing heat radiation. Special coatings are often used to increase the absorptive power of the back place.

In addition to the flat plate collector, cylindrical, double walled transparent tubing is used. A black plastic tubing has been developed which is replacing copper tubing inside the glazing. In addition to flat plate collectors, the various devices for concentrating solar energy during collection which have been developed include parabolic dishes, cylindrical parabolic reflectors, a series of flat mirrors, etc., all of which are adjusted to focus the Sun's rays on an absorber.

Storage

A second component of a solar heating system is the heat storage unit which is needed to provide for nights and cloudy days. The common storage medium is water in an insulated tank. When the heating system circulates air, however, crushed rocks or pebbles are often used. Since water has a higher heat capacity than stone (it can store more heat per pound), the storage volume of water is less.

The usual "rule of thumb" for 2 or 3 days of storage is about 60 litres of water storage per square metre of collector. If rocks are used the space provided must be larger. The rule of thumb is about 1300 kg per square metre collector.

There are promising experiments underway with more sophisticated storage techniques using solid materials (eutectic salts) which melt at about 40°C and then give up stored heat when they cool and solidify. Commercial realisation of this type of storage is still somewhere in future.

An interesting combined collector-storage system showing great promise is the "solar pond". A shallow body of water is used as the collector. Normally the bottom of the pond would absorb heat, the bottom layer of water in contact would thus be heated and through convection be transferred to the pond's surface and lost. In a solar pond, the bottom water layer is made very saline and thus is too dense to rise even when heated. A middle level of medium salinity protects the bottom layer, and the freshwater on top acts as insulation. Hot water at temperatures as high as 80° or 90°C can be obtained and stored without much loss for several days.

Efforts are now under way to develop freshwater solar ponds in which evaporation and heat loss are prevented by a honeycomb-like material on the surface which lets the Sun through while the trapped air insulates.

Solar Cooling

At first thought it seems incongruous to consider cooling with solar energy. Evaporative coolers, however, are still popular in hot, dry regions such as the southwestern United States. Water is evaporated as it runs down over a system of pipes, and the water in the pipes, cooled by the evaporation, is then circulated through the house.

Shallow rooftop ponds can be used for cooling and heating. The house is cooled by evaporation of the heated water during the day. At night, a reflective insulating cover is drawn across it, and it provides heat during the cooler evening hours.

More sophisticated systems use the "absorption–desorption" principle of the natural gas refrigerator. A refrigerator is a "heat pump". Heat is transferred by exposing an expanded and, therefore cold, gas to the cool interior; compressing it so that its temperature increases: and allowing it to discharge this heat outside. In the electrical refrigerator, pumps and compressors are used. In a heat-activated refrigerator, the heat drives a vapour out of a liquid; the vapour expands, cools, picks up heat, and is then re-absorbed in the liquid to carry the heat away. The major difficulty with solar cooling units of this type is that temperatures of 90°C or higher are needed. Since flat plate collectors cannot easily produce temperatures higher than 65°C, some kind of concentrator will be needed.

There are also plans being developed in several countries, in Israel in particular, to use the steam discharged from large central receiver thermal electric power plants, to provide the energy for a large-scale absorption–desorption cooling unit.

Solar cooling promises to make a most important energy contribution. Its major competition is expensive electricity. Since the demand for cooling usually peaks in the summer when there is plenty of Sun, solar cooling should be valuable.

Heat Pumps

There is growing interest in combinations of heat pumps and solar systems. A heat pump moves energy from one place to another. It is reversible and thus can pump heat from a storage tank (or the outside air) into a building in the winter and from the indoors out in the summer. An attractive combination is to use solar collectors to warm water in the winter and to draw from this supply with a heat pump.

The heat pump works best when the temperature difference between its source and the interior is not too great; the solar collectors work best when they do not have to provide really hot water. The efficiency of both systems is thus improved.

Solar Heating and Cooling in Practice in the United States

Both active and passive solar heating has rapidly grown in popularity in the United States. As a measure, the total area of low temperature solar collectors manufactured has increased from less than 185,000 m^2 in 1974 to greater than 740,000 m^2 in 1981. Their contribution to the total U.S. energy mix remains small however, less than 1 per cent.

The only market now dominated by solar in the United States is swimming pool heating on the West Coast. With the cool nights (and often cool days) there, many residents in the past heated their pools with (then) inexpensive natural gas. They now use floating metal or plastic collectors which transmit the heat absorbed to the pool. The manufacturer of solar swimming pool heaters has become an important industry.

Resources

A major advantage of solar energy is the amount of this resource. The rate at which solar energy falls on the upper atmosphere (the power input) is 1.36 kW per m^2. About a third of this energy is immediately reflected away. The amount which arrives at the ground depends critically on latitude, the season, local weather conditions, etc., but the day/night U.S. average is 177 W per m^2 or about 13 per cent of the original power.

This is not a lot of power; it would take almost a metre square collector to provide enough power for a 200-watt bulb. It adds up, however; in 24 hours this is a little more than 15MJ/m^2. The energy equivalent of ten barrels of oil falls on each U.S. acre per day, and the equivalent of 22.4 billion barrels of oil per day on the 2,310 million acres of the forty-eight contiguous U.S. states (almost four times the U.S. present yearly oil consumption).

Solar resource for the United States

Solar input varies considerably from the north to the south in the United States and to a lesser extent from winter to summer. In January for example, the northernmost U.S. regions get between 1000 and 2000 Kcal/m^2 daily while the deep south can receive as much as 3000 Kcal/m^2. In July the energy input is much more consistent (as clouds are not as important) and ranges between 5000 Kcal/m^2 in the far north to 7000 Kcal/m^2 in the south.

Environmental Considerations

One of the perceived advantages of solar energy is its low environmental impact. Sunlight falls whether we use it or not. Thus there is no pollution to discharge. In a heavily solar-dependent urban area there may develop problems over solar access, shading, etc., as well as aesthetic consequences. The major environmental impact is apt to be indirect. A lot of conventional energy with its associated environmental impact will be required if a nation were to undertake the manufacture of a large number of active solar systems.

Summary

Solar energy for heating and cooling has the advantages of a large resource base and widespread distribution. Its disadvantages are that it is spread rather thinly and thus must be collected over relatively large areas, and is erratic, fluctuating day to night and with weather conditions.

Solar energy is a "free" fuel, but since it requires such a large collection area it often requires a large capital investment to make use of it. It does not yet, in many countries compete with inexpensive commercial fuels or with easier to gather non-commercial ones. Solar energy, however, is the only large resource among the renewable fuels and will steadily become more important as the technology of collection, storage, and use advances, and as other sources of heat become more scarce and expensive.

30

Conventional Nuclear Reactors

J. M. FOWLER

NSTA, Washington, U.S.A.

Most of the commercial and non-commercial energy used in the world has been stored as chemical energy. It depends basically, as does all chemistry, on rearrangements of electrons in atoms and molecules. Energy is stored by acting against a force. Chemical energy is stored by acting against the electrical force. The amount which can be stored per atom or molecule therefore is dependent on the strength of the electrical force.

The only exceptions to the dominance of chemical energy in human service until the 1940s were wind and water power. These forms of energy involved the much weaker gravitational force and thus the energy density in lifted water and in blowing wind is much smaller than that in coal or wood.

With the discovery of nuclear fission in the late 1930s and the development of techniques to control it in the 1940s a new source of energy was added to the menu. Nuclear energy is released by rearranging the neutrons and protons which make up the nucleus, the tiny dense core of the atom. The energy density in the nucleus, the energy per atom which can be released, depends on the strength of the nuclear force. Since this force is about a million times larger than the electrical force which characterises chemical energy, the energy per atom obtainable in a fission reaction is about a million times more than is available from, for example, burning coal.

This speculation is borne out in practice. The complete fissioning of a pound of uranium could produce nine billion kilocalories of energy, about 1.5 million times more energy than is released by burning a kilogram of coal.

It is no wonder that humankind has looked to this "wonder source" with eager hope for four decades. Unfortunately nuclear energy from fission, as it has developed, has followed the first law of energy technology (sometimes known as "Fowler's Law"): "The farther from practical use a technology is, the better it looks." Nuclear energy, once seen as producing electricity which would be "too cheap to meter" now shows a face of warts and scars.

Technology

"Converter reactors" as distinguished from "breeder reactors" obtain their energy from the fissioning of a rare form of uranium, uranium 235. U-235, or uranium 235, is the accepted symbol for the isotope of uranium which has 235 particles in its nucleus. It is thus distinguished from U-238, another isotope, chemically identical to U-235, but possessing different nuclear properties. This isotope makes up less than 1 per cent (actually 0.7 per cent) of common uranium ore, most of which is U-238. U-235 releases energy when the nucleus undergoes fission, that is, splits into two fragments.

Fission is induced by a neutron, an uncharged particle, which together with the proton constitute the building blocks of the nucleus. When a neutron strikes a U-235 nucleus, energy is released causing the nucleus to become unstable and finally to break into two pieces with the release of much more energy.

In the fissioning of U-235 several more neutrons are released—on the average about 2.5. These neutrons can then strike other U-235 nuclei causing them to fission. Since the neutrons move very fast, the time between successive generations of fission events is very short, about 0.1 ms. This geometrical increase goes very fast and results in a "chain reaction".

Nuclear Reactors

The purpose of a nuclear converter reactor is to control the chain reaction and extract energy from it. The reaction fuel, a mixture of the fissile isotope U-235 and the more common isotope, U-238, is packed into thin rods, about 1 cm in diameter and about 3 m long which are assembled into a "core". This core is surrounded by shielding which helps turn some neutrons, which may have escaped, back into the reacting region. Most forms of reactors are filled with a "moderator", a light medium such as graphite, water, or deuterium oxide—heavy water. The moderator slows the neutrons from the fissioning nuclei down to a speed at which they have a good chance of causing other fissions.

To the core and the moderator must be added control devices to keep the chain reaction from producing excessive heat. These rods contain neutron absorbing materials such as cadmium and boron. The rods are moved in or out of the core to control the number of neutrons available for fission and in that way control the power output of the reactor.

The Energy Conversions

When fission occurs, the released energy appears as kinetic energy of the particles. This kinetic energy is quickly transformed to heat energy via the collision process. Thus the nuclear reactor is a source of very high temperature heat energy.

Different reactor types extract this heat energy in different ways, but basically it is used to boil water to make steam. The steam turns a turbine-generator system to create electricity in much the same manner as does steam in a coal fired plant. A nuclear power plant is a heat engine. Only the source of heat is unusual.

The Fuel Cycle

All power plants have an associated fuel cycle. Coal must be mixed, prepared, delivered to the power plant, and its ashes must be disposed of. The fossil-fuel "fuel cycles" are so familiar that we pay little attention to them. (Although coal mining has environmental impact as does ocean transport of oil.) Because the nuclear fuel is radioactive and requires energy-expensive processing, however, it deserves special attention.

We describe briefly, below, the step-by-step movement of uranium from mine to reactor, etc. Some additional technological details will be developed in the section on Safety.

The Fuel Cycle

Since the raw ore only contains a small percentage (2 or 3 per cent) of uranium, 100,000 or more tons of ore must be mined to provide the 150–200 tons of uranium oxide, U_3O_8, which is about the annual requirement of a 1000-MW nuclear reactor. The ore is refined and concentrated in the milling operation. It then goes through a conversion to a gas, uranium hexafluoride, which is in the form in which enrichment takes place. In the enrichment plant the amount of U-235 is increased from the original 0.7 per cent to 3 or 4 per cent. This enriched uranium is then assembled into the fuel elements.

After a year's operation the fuel is removed. The present practice is to replace one-third of the reactor fuel each year. The initial loading of the 1000-MW reactor would therefore take three times the amounts of uranium oxide as is consumed in a year. The reprocessing steps are not presently in operation but may eventually return uranium and plutonium to the fuel cycle. The other flow out of the reprocessing plant is of radioactive waste materials which require long-term storage. We will refer back to this fuel cycle in our later comments on safety and waste disposal.

Resources

Uranium occurs in small amounts in most rock. To be mined, however, the deposit must be concentrated. Commercial ore, on the average, contains 4 or 5 pounds of uranium oxide (the commonly occurring uranium compound per ton). Of these 4 or 5 pounds, only 0.02 to 0.03 pounds are the sought-after U-235.

Uranium fuel production begins with a search for ore that is rich enough to mine and a lot of rock and dirt is moved in the process. The raw ore is then "milled", concentrated to a product called "yellowcake" which is 80 per cent uranium oxide. If a 1000-MW reactor were to operate at 70 per cent efficiency, it would use 130 tons of yellowcake per year.

Resource estimates are given in terms of tons of yellowcake and are classified by the cost in dollars per pound in producing it. At present, reactors use reserves costing no more than 50 dollars per pound. As of 1982 there were 594,000 tons of U.S. reserves in this price range and perhaps 2,000,000 additional tons known to exist worldwide. Since exploration and analysis is still very incomplete worldwide, much more low cost ore may still be found.

To put the resources in energy terms we need to use the information previously given in the fuel cycle description. A 1000-MW reactor operating at 70 per cent efficiency uses 130 tons of yellowcake per year. Efficiency here means that it produces 70 per cent of the rated energy of the plant. This is also called the "capacity factor". Producing 700 MW of electrical power (70 per cent of 1000 MW) for a year means the production of about six billion kWh of electrical power.

A more interesting example is to calculate the amount of generating capacity that the uranium reserves can support. Over its expected thirty-year lifetime, a 1000-MW nuclear power plant will use 30 × 130 tons or about 4000 tons of uranium oxide. (This is probably an underestimate.) Thus, for instance, the 954,000 tons of uranium oxide, U.S. uranium reserves at under 50 dollars per pound identified in 1982 would fuel 236,000 MW of generating capacity or 236 of the 1000-MW reactors. (A similar calculation can be done for other countries with reactor programmes.)

It is clear in the United States that either more inexpensive uranium ore will have to be discovered or the breeder reactor, which can use the more abundant U-238 will have to be implemented.

Environmental Considerations

Nuclear reactors, if they come to dominate electric power production, will ease several of our perplexing environmental problems. They do not produce the air pollutants presently associated with coal or other fossil-fuel plants. The amount of ore to be mined, 100,000 tons of uranium vs. 2,500,000 tons of coal per year for a 1000-MW plant, would help to reduce the worry over strip mining. Reactors, however, have their own environmental impact.

We will discuss two types. The more or less "conventional" pollution problems posed by the radioactivity which is a by-product of the fission process and a broader set of hazards which we will deal with under the title Safety.

Radioactive Pollution

The fission fragments are radioactive. They emit energetic particles or "radiation" which can penetrate tissue or other matter. The amount of radioactive material which accumulates during a year's operation of a reactor could have a large impact on the health and safety of the general population. Radioactive pollution is, of course, rigorously controlled at the reactors. The regulatory guidelines are quite strict and a person could, in fact, quite safely (as far as routine emission of radioactivity is concerned) live right at the plant boundary line. The exposure caused by the routine operation of even the several hundred reactors projected for the future will still be small compared to other sources of radiation exposure.

Radioactive Waste

The left-overs, the "ashes" of the conversion process, are not only radioactive, but remain so for a long time. The crucial parameter of the radioactive substance is its "half-life", the time during which half of a sample of the material will "decay", or change into something else. Radioactive species with short half-lives are intensely dangerous for short times, while longer half-lives mean that the radiation is less intense, but is spread out over longer time.

In assessing the danger to man, one has to consider, in addition to the half-life, the biological properties of a contaminant. Several of the fission products have a high biological impact in addition to a long half-life so must be kept completely out of the ecosystem for hundreds of years. It is this necessity for complete isolation that is the challenge of radioactive waste disposal and which makes the discussion so often acrimonious. Questions are raised which are both ethical and technological. We are, in effect, leaving this material to our descendants. It will have to be stored and monitored for several generations.

The technological solution seems to be easier than the ethical. One plan being considered calls for the flash drying of the high level liquid waste to a ceramic material with a large reduction in volume. These "hot rocks" would then be stored for at least a few hundred years in some dry and geologically isolated spot, such as a salt mine for instance.

The weakness of this technological solution, constantly pointed out by the nuclear critics, is that no agreed-to official plan presently exists, and the waste is not as yet being processed in this way. It is piling up rapidly in temporary storage. The resolution of this waste disposal problem must come quickly if the reactor programme is to grow at its projected rate. The problem must be faced, however, even without additional nuclear power because of the waste already generated in weapons production.

Safety

The most emotion charged issue in the nuclear controversy is the safety issue. There are actually three parts to the safety question: the safety of the transport of radioactive fuels and waste material, the safety of nuclear fuels from the danger of hijacking by would-be bomb makers, and the safety of nuclear plants themselves.

We can deal with the first part rather swiftly. Although there is a growing concern over the increasing volume of radioactive materials on our roads and railways (fifty-two shipments per day in 1980 in the United States), the great care exercised so far has prevented serious accidents and this problem seems manageable.

The danger from hijackers seem more threatening, and it is a danger that may grow if we institute plutonium recycling and the breeder programme.

The focus of much anti-nuclear concern is on the danger of an accident at the reactor which will release some of its radioactive products to its surroundings. It is not a nuclear explosion which is feared. U-235 is not present in sufficient concentration to allow a runaway chain reaction to occur. The greatest danger is that a reactor will, for some reason, lose its coolant, melt down, and either melt through its containment or fracture it by a steam explosion. If this happens, some of the radioactive by-products may be released to the environment.

The fission reaction which produces the energy in the nuclear power plant is a self-sustaining one. We have described the reactor control mechanism. If a reactor malfunctions, it will be shut down, or "scrammed", by running all the control rods into the core.

The reactor remains very hot for some time after shutdown, and cooling is thus extremely important. Without a cooling system, a core would melt within several minutes. To prevent this from happening, there is, in addition to the regular cooling system (the water which carries the heat energy to the steam generator), a back-up "emergency core cooling system" designed to dump an enormous amount of water into the reactor if the regular system fails. Much of the controversy over reactor safety has focused on this emergency system which has never been fully tested.

There have been accidents however, the most publicised were the Three Mile Island, TMI, reactor in the United States in March of 1979 (producing no deaths but caused a tightening of regulatory controls and safety design improvements in the United States) and the Chernobyl reactor in the Soviet Ukraine in April of 1986 (causing several deaths and as yet unknown long-term health and cancer problems). Since those reactor accidents were caused by human error and equipment malfunction, it has greatly undermined the public's confidence in reactors. In addition, the great cost of the failures, running to several billions of U.S. dollars, has understandably undermined investor confidence.

Summary

The irresistible analogy that springs to mind when discussing nuclear energy is the genie released from the lamp. It is a genie which offers great power at a time when our traditional sources of energy and power are disappearing. Yet it is a genie needing rigorous control.

Nuclear fission, if the breeder cycle is implemented, is one of the "big three" (along with solar and nuclear fusion) energy sources which are not resource-limited. It produces electricity, a most convenient and desired form of energy. Currently, it is only used to produce electricity. However, the world needs portable energy for transportation and heat for its buildings.

Many countries are already committed to nuclear reactors but many more have adopted a wait and see attitude. The wait will not be long, and during the next decade or two we should begin to see the future role of nuclear reactors.

31

Wind Power

J. M. FOWLER

NSTA, Washington, U.S.A.

For centuries, power from the wind has been used to pump water for crop irrigation, to propel sailing ships across the ocean, and to turn millstones to grind flour from grain. From its origins in Persia in the seventh century, the windmill concept spread throughout the Islamic world and reached Iran and China by the time of the Mongol conquests. In the seventeenth century, the Netherlands had become the world's most industrialised nation by extensive use of wind power to drive trading ships, to power grinding mills, and to pump water from lands that were once beneath the sea.

In the mid-nineteenth century, multivane windmills were used in both Australia and the United States to pump water on the large cattle ranches. Small-scale generators of a few kilowatts were used to generate electricity for isolated farms and homesteads in the United States and Europe until they were replaced by electrification programmes.

Interest in the potential of large-scale wind generators, increased between 1935 and 1955, and a number of machines, ranging in power from 90 kW to 1125 MW, were built. One of the largest and most well known, the 1.25 MW Smith-Putnam generator, was built on "Grandpa's Knob" in the mountains of central Vermont. While the success of that experiment was short-lived, due to mechanical failure and the complications and expense of war-time materials supply, it did demonstrate the feasibility of capturing wind power for electrical generation on a large scale.

Similar large machines were built in Europe. The Gedsee windmill in Denmark, a 200-kW, 25-m diameter machine built in 1957, operated successfully for eight years; a 100-kW, 35-m diameter machine, built in 1957 also, at Stohen in West Germany, operated for seven years.

Wind power differs from most other "alternative energy technologies" in that it is an old and proven source. Its demise was brought about to some degree by the overly optimistic projections of cheap nuclear power and the unrealistic energy prices of the 50s and early 60s. It is a technology which could be quickly brought to a large-scale use given determination and support on the part of the private and public sectors.

Technology

The wind gets its energy from solar heating of the Earth's surface. The energy is the kinetic energy of mass in motion. The laws of physics tell us that this kinetic energy varies as the square of the wind speed. Since the amount of air which passes through the diameter of a windmill per unit time depends on the first power of the velocity of the air, the power available depends on the third power or cube of the wind velocity. Thus, if the wind speed doubles, the power available increases eight-fold. It is thus important for wind machines to operate at as high a wind speed as possible.

Since the energy comes from the cylindrical volume of air which passes through the blades, the power available varies as the square of the windmill's diameter. Thus if the diameter is doubled, the power output is quadrupled (assuming the instantaneous wind velocity is the same throughout this cylindrical volume of air).

A final consideration of importance is that the wind speed increases with height above the Earth's surface, at least within the first hundred metres. An increase of 25 per cent between 10 m and 100 m is more or less typical. Thus the machines designed for large power output have large blades sitting on tall towers. Practical considerations, however, work against tall towers. The wind exerts a large leverage on the tower, tending to tip it over.

No wind turbine can operate at or even near 100 per cent. If it could extract all the energy from the moving air, the air would stop and obstruct that following it. It can be shown that the upper limit on the efficiency of a wind machine is about 59 per cent. This is known as the Betz factor. Some of the large propeller-driven machines are nearing that limit.

Vertical and Horizontal Axis Machines

There are basically two types of wind machines, horizontal axis wind turbines, HAWT, in which the turbines and usually the generators are mounted at the top of the towers, and the vertical axis wind turbine, VAWT, in which the generator can be housed at ground level. The HAWT needs to be turned to face the wind while the VAWT, of which the most familiar example in the eggbeater shaped Darrieus rotor, can accept wind from any direction.

Wind turbines are not "pushed" by the wind as water wheels are by water, for then their speed could not exceed the wind speed. They are driven by the aerodynamic lift force of the type which keeps airplanes aloft. Wind flowing over a properly designed blade creates a pressure difference which drives it. The blade tips can move at speeds much higher than the wind.

The design of any wind machine depends on the task assigned it. If it is to pump water it needs a large starting torque. The multivane machines with many blades to catch the wind provide this torque. If electrical generation is the task, then high-speed turbines are needed because the voltage generated depends on the rotational speed of the generators.

The VAWTs are generally smaller machines used principally for electrical generation. They are designed for high speed and most need to be started with the generator reversed and acting as a motor.

Storage and Interconnection

Wind power shares the inconsistency of most other solar sources. Even in areas of high wind intensity there are daily and seasonal variations. There are in addition minimum and maximum wind speeds for the operation of any specific machine. Thus the "capacity factor", the percentage of time during which a machine produces its rated power, is only 20–25 per cent.

If the machine is used for an interruptible task, pumping water for instance, the intermittency is no problem. For electrical generation, however, this intermittency of energy delivery has to be compensated for by storage and/or interconnection. For a grid of machines spreading over a large geographical area the variation can be, to an extent, smoothed out. The wind is usually blowing somewhere.

Wind also has a natural complementarity with its solar cousins, water and sunlight. To a degree, the wind tends to blow when the Sun is not shining and vice versa. Some complementarity is expected on a day/night basis and some seasonally. The winds, in particular, are stronger in the winter when the solar input is apt to be lower. Solar and wind power electrical generation would tend to be more even than either alone. In areas where hydropower is used, wind power can be made use of when it is available, letting water build up behind the dam to supplement the wind when it dies.

Wind power, however, will probably always provide supplemental power, perhaps contributing 25 per cent, to a grid dominated by fossil fuel or nuclear powered generation.

Wind Power in the United States

Wind power in the United States has developed along two lines. Three big HAWTs with 91-m rotors each capable of generating 2.5 MW of electrical power (in a 27 mph wind) have gone into service along the Columbia River gorge near Goodrise Hills in the state of Washington; other large machines are planned in the west. There are many smaller machines elsewhere at windy sites throughout the country. The tax incentives and environmental restrictions in California make wind power especially attractive there and "wind farms" are being rapidly established in the windy passes of the California mountains. It is expected that the wind powered generating capacity of these "farms", which sell their electricity to the large utilities, will reach 700 MW by 1987.

Resources

Wind energy is solar in origin and thus can provide no more power than the 180 TW solar flux to Earth. Most of this flux is, of course, not available as wind. To set an upper limit on the wind resource we must look only at the solar energy fed into the Earth's atmosphere by surface heating and settle for only a small fraction of that. A global upper limit of 130 GW has been suggested. This is ten times the total world energy consumption.

The practical upper limit is even much smaller. We resist the idea of covering our land with wind machines. Thus the present approach is to concentrate on isolated, high mean windspeed sites. Islands off the coast are particularily attractive for siting wind machines as they replace costly diesel-fueled generators.

Environmental Considerations

Because wind machines are relatively passive "low technology" machines, there are no direct by-products or residues to dispose of, and the secondary effects of their manufacture are not especially troublesome.

Their major drawback is land use. Although a single machine does not occupy much space, a giant grid of them complete with power lines would have aesthetic drawbacks, but wind harvesting should be compatible with other uses of the land such as pasturage and farming.

The efficiency of land use is also expected to be comparable to farm use. If "solar efficiency" is defined (for a large area windmill farm) as the ratio of power output from the windmills to the solar power incident on the land they occupy, it is estimated that efficiencies of about 5 per cent are possible in the great plains area. Photosynthetic efficiency of growing plants there is about 1 per cent.

If really large-scale use becomes a reality, there will be questions to answer about possible effects on weather caused by the removal of large amounts of kinetic energy. In general, however, wind energy, like other forms of solar energy, seems to be environmentally benign.

There are other considerations. New wind machines may not be considered to be as beautiful as the Dutch windmills. Some of them have been noisy, and if the blades are of metal they interfere with television reception. But since they are necessarily removed by 200 m or so from habitation so that broken blades do no damage, noise and TV interference will be minimised.

Summary

The wind is always with us and we have sufficient technological know-how to tap it. No one questions that the technology can be improved and that there will be problems to overcome. The task of building a large grid of wind machines is, however, one that could be accomplished relatively inexpensively and, in contrast to other power plants, quickly. (The 1.25 MW generator on Grandpa's Knob was constructed in nineteen months.) If the priority and support is forthcoming, we can begin to define the windmill's new role before the decade of the 80s is over.

32

Geothermal Energy

J. M. FOWLER

NSTA, Washington, U.S.A.

Geothermal energy, the natural heat of the Earth, promises to be an important source of supplemental power in selected areas around the world. It is estimated that the Earth's geothermal heat may represent a total energy resource equal to 400 million billion barrels of oil per square kilometre in the top 10 km of the Earth's surface. While most of this energy is too diffuse to be recovered, there are concentrated pockets of heat within the Earth's crust that are accessible and that could provide a significant contribution to local energy needs.

Geothermal energy is primarily the product of a few elements in the Earth's core which release energy as they radioactively decay. This energy is then converted to heat energy that keeps the core molten. In certain places geological forces thrust large pools of this molten rock relatively near the Earth's crust and these form the so-called geothermal "hot spots". These hot spots occur most frequently near the junctures of the continental plates, where they are colliding or drifting apart. It is also true that high surface heat flows and the welling up of magma occur.

Geothermal systems can be divided into four catagories: hydrothermal systems (hot water and steam), dry hot rock, magma, and geopressured systems. There is also the "normal gradient". As one drills deeper into the Earth's crust the temperature rises, on the average about 30°C per km. There is thus an enormous amount of low-quality heat energy available at reasonable depth below the surface.

The hydrothermal systems depend on magma being thrust up near to the surface. These systems and the pure magma system (volcanoes) are most common along the continental plate junctures. The hot dry rock systems are more widely distributed.

Geopressured systems are quite different. Typically they are fossil water that has been trapped under an insulating and impermeable layer of clay. Energy of three types is available. The trapped water is heated by conduction to relatively high temperatures (200°C) and is also at high pressure, as high as 10,000 lb/in^2. If, as is often the case, organic matter was trapped in them, it has been converted to methane. The geopressured systems are known to be extensive but are now of little commercial interest.

Neither the dry hot rock systems nor the magma systems are presently important although the former are widespread and the latter are tremendously concentrated local sources. We have not yet mastered the technology of drilling into the hot rocks, creating a reservoir, and circulating water to recover the heat; nor do we have equipment sufficiently heat resistant to deal directly with the molten magma in volcanoes.

Technology of Extraction and Conversion

We have practical experience only with the hydrothermal systems. These systems can be mostly steam at temperatures greater than 180°C; they can be a mixture of steam and water; or they can be hot water. To be of use in the generation of electricity a temperature greater than 80°C is generally required.

The vapour dominated or "dry stream" fields (those which contain mostly steam) are the most commercially attractive, because the steam can be used, as is, to power turbines. The only U.S. field, the Geysers in California, is of this type, as is, the Lardello field in Italy. The hot water or "wet steam" fields are at a somewhat lower temperature and contain a mixture of steam and hot water. They are more difficult to work with since the steam must be clear of water for use in a turbine. These are about twenty times more common than the "dry steam" fields.

The technology for producing electricity from hot (greater than 210°C) steam is essentially the same as for other steam electric production. Wells are drilled, and the steam from several wells is sent through insulated pipes to a turbine-generator combination. Filters to clean the steam of sand and small rocks are provided. Since the turbines operate at lower temperatures and pressures than in fossil-fuel plants the efficiencies are lower; about 15–20 per cent is usual.

Wet steam can be used in a similar fashion by "flashing" some of the superheated water (water under pressure at a temperature above its usual boiling point) to steam. The steam is separated from the water and used.

At temperatures less than 100°C, so-called "binary" cycles must be used. The hot liquid from the geothermal reservoir is pumped into a heat exchanger and heats another liquid, like freon or ammonia, which has a lower boiling point. This liquid is vaporised and the vapour turns a large, low speed turbine. The Israeli turbine-generator is an example of such a device. The geothermal fluid is returned to the well. The vapour is condensed to a liquid by a cooling system as in any heat engine.

Geothermal Energy in Practice

In the early use of geothermal sources, health rather than energy was the goal, and hot springs were popular gathering places. Italy was the first country to generate electric power with natural steam by opening the Lardarello Field in 1904. This dry steam field now has 396 MW of generating capacity. Other developed fields exist in New Zealand (where wet steam fields produce not only 170 MW of electric power, but heat for homes and industry as well); in Japan (50 MW); in Mexico, which has just installed a 75-MW generator at Cerro Prieto in the north; in Russia, and in Iceland. Mexico, Japan, the Central American Countries, and the Soviet Union all have ambitious plans for expansion.

Use of geothermal heat directly is also of increasing importance. In Iceland geothermal heat has been used for a long time in buildings; it provides almost all of the heat for the 100,000 people in the capital, Reykjavik. A 60-MW electric power plant is under construction in northern Iceland. Several hundred American homes and businesses are heated in this way in Klamath Falls, Oregon, and Boise, Idaho.

Geothermal home heating is also used in Budapest, Hungary, and use is growing; the goal is 100,000 to 200,000 dwelling units. There are some 20,000 geothermally heated apartments outside of Paris, France.

Geothermal Energy in the United States

While some direct heating of houses and industry is accomplished in the western United States, the most important U.S. geothermal conversion is in the Geysers near San Francisco.

By 1983, seventeen turbine-generators were generating 1100 MW of electric energy, about one-third the electric power needs of San Francisco. Extensive drilling continues, and an additional 100 MW of generating capacity per year, toward a total of perhaps 6000 MW by the year 2000, is planned. Extensive exploration both for steam and hot water continues in the west.

Resources

Most of the presently developed geothermal resources were discovered through obvious surface indications: geysers, steam from fissures, pools of hot brine. More sophisticated techniques for exploration, infra-red surveys by satellite, for instance, are now available and will allow more meaningful resource estimates in the future.

Geothermal hot spots are most likely to occur in the volcanically active, earthquake-prone regions of the Earth's crust: down the Western coast of the Americas from Alaska to Chile, down the Rift Valley into Africa and across into Turkey, and in the Far East along the "Circle of Fire" that borders the

Pacific. Some countries have especially large resources: two-thirds of Turkey is believed to have geothermal potential, and one U.N. survey suggests that the geothermal potential of Ethiopia if developed could satisfy the present electrical needs of Africa. The heat is there; imagination, money, and determination are needed to bring it into service.

National Resources in the United States

A preliminary survey and analysis of U.S. geothermal resources by the National Research Council gives an estimate of about 23,000 quads. The energy content of geopressured brines, normal gradient (the heat found beneath the surface due to conduction from the core) and that associated with hot dry rock and magma accounts for about 74 per cent of that total and is of course quite speculative. It will not become a realistic factor until new extraction and conversion techniques are developed.

Environmental Considerations

One of the strongest arguments for the increased exploitation of geothermal power is that there are very few environmental problems associated with it. The most serious potential problem is that boron and other chemicals (which are mixed in with the geothermal water and steam) can cause damage if they are released into nearby waters or allowed to escape into the atmosphere. It is hoped that this problem can be solved by simply re-injecting these chemicals back into the well. The other major hazard is that of land subsidence from the withdrawal of large amounts of water, but again, it is thought that this can be controlled by limiting withdrawal to a safe rate and by re-injecting water into the wells, a practice currently used in many oil drilling operations.

There is some worry that earthquakes may be triggered by re-injected water under pressure. Less significant but still "nuisance" problems have to do with the odour of hydrogen sulphide which is released during exploration (the familiar rotten egg smell). This can be a serious health hazard, particularly to drilling crews, if allowed to reach high concentrations. There is also considerable noise involved as steam pushed out of the well at supersonic speeds, often exceeding 100 decibels when a well is vented. Finally, large land areas are required. This amounts to between 3000 and 5000 acres for a 1000-MW plant, with about 7–10 per cent of this used directly for facilities, a large part of which are an unsightly jumble of pipes. However, the geothermal generating plants appear to present less environmental threat than either nuclear of fossil-fuel plants, and the problems that do exist are highly localised and seem to have available solutions.

Summary

Geothermal energy is a good example of an "appropriate technology". Where it exists in shallow deposits of steam or hot water it is an excellent source of relatively inexpensive energy. It is being increasingly developed in both direct heating and electricity generation around the world. Because steam and hot water cannot be piped very far without excessive energy loss, geothermal heat will have to be used locally. This restriction to local use and its limited distribution gives geothermal a minor role in the worldwide energy mix. Where it exists, however, it has the potential for an important local role.

Index